The Book of Universes

Theories of Everything

The Left Hand of Creation
(with Joseph Silk)

L'Homme et le Cosmos
(with Frank J. Tipler)

The Anthropic Cosmological Principle
(with Frank J. Tipler)

The World within the World

The Artful Universe

Pi in the Sky

Perché il mondo è matematico?

Impossibility

The Origin of the Universe

Between Inner Space and Outer Space

The Universe that Discovered Itself

The Book of Nothing

The Constants of Nature:
From Alpha to Omega

The Infinite Book:
A Short Guide to the Boundless,
Timeless and Endless

New Theories of Everything

Cosmic Imagery:
Key Images in the History of Science

100 Essential Things
You Didn't Know You Didn't Know

The Book of Universes

John D. Barrow

THE BODLEY HEAD
LONDON

Published by The Bodley Head 2011

2 4 6 8 10 9 7 5 3 1

Copyright © John D. Barrow 2011

First published in Great Britain in 2011 by
The Bodley Head
Random House, 20 Vauxhall Bridge Road,
London SW1V 2SA

www.bodleyhead.co.uk
www.rbooks.co.uk

Addresses for companies within The Random House Group Limited can be
found at:
www.randomhouse.co.uk/offices.htm

The Random House Group Limited Reg. No. 954009

A CIP catalogue record for this book
is available from the British Library

ISBN 9781847920980

Mixed Sources
Product group from well-managed
forests and other controlled sources
www.fsc.org Cert no. TT-COC-2139
© 1996 Forest Stewardship Council
FSC

Typeset by Palimpsest Book Production Limited, Falkirk, Stirlingshire
Printed and bound in Great Britain by
Clays Ltd, St Ives PLC

To Tilly

For whom there are many universes to come

Contents

Something unknown is doing we don't know what.

Arthur S. Eddington

Preface

Universes are big at the moment. And this is a book about universes. It is a story that revolves around a single unusual and unappreciated fact: that Albert Einstein showed us how to describe possible universes – entire universes. Previously, there had been discussion about the structure of our universe for thousands of years. All manner of exotic pictures had been created to describe or explain it. The results were often driven by religious, nationalistic, artistic or personal prejudices. They were stories. In the early twentieth century, things suddenly changed: Einstein showed us how to find all the possible universes that were consistent with the laws of physics and the character of gravity, how to reconstruct their pasts and predict their futures. But actually finding them was no easy task. Ever since, astronomers, mathematicians and physicists have struggled to solve Einstein's intricate equations and find these universes. This book is about that struggle and the possibilities it gradually uncovered.

What a gallery of possible universes there are. Many bear the names of some of the most famous scientists of the twentieth century. Some expand in size, others contract, some rotate like a top while others are totally chaotic. Some are perfectly smooth, while others are lumpy, or shaken in different directions by cosmic tides; some oscillate forever, some become lifeless and cold, while others head towards a runaway future of ever-increasing expansion. Some permit time travel into the past, while others threaten to allow infinitely many things to happen in a finite amount of time. Only a few allow life to evolve within them; the rest must remain unexperienced by conscious minds. Some end with a bang, some with a whimper. Some don't end at all.

Our story will encounter universes where the laws of physics can change from time to time and from one region to another, universes that have extra hidden dimensions of space and time, universes that are eternal, universes that live inside black holes, universes that end without warning, colliding universes, inflationary universes, and universes that come into being from something else – or even from nothing at all.

Gradually, we will find ourselves meeting the latest and the best descriptions of the universe we see around us today, together with the concept of the 'Multiverse' – the Universe of all possible universes – that modern theories of physics lead us to contemplate. These are the most fantastic and far-reaching speculations in the whole of science. They challenge us to ask whether the exhibits in our gallery of possible universes actually exist or whether there is only one that achieves this special status.

Other cosmology and astronomy books have focused on particular topics – dark matter, dark energy, the beginning of the universe, inflation, life-supporting coincidences, or the end of the universe – but this book introduces the reader to whole universes and the histories of their discovery, along with the personalities of the scientists who found them, in a coherent and unified way.

I would especially like to thank Katherine Ailes, Allen Attard, Donato Bini, Arthur Chernin, Hyong Choi, Pamela Contractor, Cecile De Witt, Charles Dyer, Ken Ford, Carl Freytag, Gary Gibbons, Owen Gingerich, Jörg Hensgen, Bob Jantzen, Andre Linde, Kay Peddle, Arno Penzias, Remo Ruffini, Doug Shaw, Will Sulkin, Kip Thorne and Don York for their help with editing, gathering pictures and providing other important historical details. I would also like to thank Elizabeth for her careful support and our now not-so-young children for their questions and the granddaughter to whom this book is dedicated.

John D. Barrow
Cambridge

1 Being in the Right Place at the Right Time

I know it's all in our minds, but a mind is a powerful thing.
  Colin Cotterill[1]

TWO MEN WALKING

I am always surprised when a young man tells me he wants to
work at cosmology; I think of cosmology as something that
happens to one, not something one can choose.
  William H. McCrea[2]

The old gentleman walking down the street looked the same as ever
– distinguished but slightly dishevelled, in a Bohemian style, a slow-
walking European on an American main street, sad-faced, purposeful
but not quite watching where he was going, always catching the attention of the locals as he made his way politely through the shoppers and
the contra-flow of students late for lectures. Everyone seemed to know
who he was but he avoided everyone's gaze. Today, he had a new
companion, very tall and stockily built, a little the worse for wear, untidy
but in a different way to his companion. They were both deep in
conversation as they made their way, walking and talking, oblivious of
the shop windows beside them. The older man listened thoughtfully,
sometimes frowning gently; his younger companion enthusiastically
pressing his point, occasionally gesticulating wildly, talking incessantly.
Neither spoke native English but their accents were quite different,
revealing resonances with many places. Intent on crossing the street, they
stopped, lingering at the kerbside as the traffic passed. The traffic lights
changed and they continued quietly across the street, both momentarily

concentrating on light, sound and relative motion. Suddenly, something happened. The taller man started to say something again, making a dart of his hand. The traffic was moving again now but the old man had stopped, dead in his tracks, oblivious to the cars and the hurrying pedestrians. His companion's words had consumed his thoughts entirely. The cars roared past on both sides leaving the two of them marooned in their midst like a human traffic island. The old man was deep in thought, the younger one reiterating his point. Eventually, resuming contact with the moving world around them, but forgetful of where they had been going, the older man led them silently towards the pavement – the one they had stepped off a minute ago – and they walked and talked their way from whence they had come, lost in this new thought.

The two men had been talking about universes.[3] The place was Princeton, New Jersey, and the time was during the Second World War. The younger man was George Gamow, or 'Gee-Gee' to his friends, a Russian émigré to the United States. The older man was Albert Einstein. Einstein had spent the previous thirty years showing how we could understand the behaviour of whole universes with simple maths. Gamow saw that those universes must have had a past that was unimaginably different to the present. What had stopped them both in their tracks was Gamow's suggestion that the laws of physics could describe something being created out of nothing. It could be a single star; but it might be an entire universe!

FUNNY THINGS, UNIVERSES

> History is the sum total of the things that could have been avoided.
>   Konrad Adenauer

What is the universe? Where did it come from? Where is it heading? These questions sound simple but they are amongst the most far-reaching that have ever been posed. Depending upon how much you know, there are many answers to the question of what we mean by 'universe'.[4] Is it just everything you can see out in space – perhaps with the space in-between thrown in for good measure? Or is it everything that physically exists? When you draw up the list of all those things to include in 'everything' you start to wonder about those 'things' that the physicists call the 'laws of Nature' and other intangibles like space and time. Although you can't touch or see them, you can feel their effects, they seem pretty important and they seem to exist – a bit like the rules of football – and we had better throw them in as well. And what about the future and the past? Just focusing on what exists *now* seems a bit exclusive. And if we include everything that has *ever* existed as part of the universe, why not include the future as well? This seems to leave us with the definition that the universe is everything that has existed, does exist and will ever exist.

If we were feeling really pedantic we might take an even grander view of the universe, which includes not only everything that can exist but also everything that could exist – and finally, even that which cannot exist. Some medieval philosophers[5] were drawn to this sort of completeness, adding everything that has existed, does exist and will not exist to the catalogue of what was, is and will be. This approach seems bent upon creating new problems in an area where there are enough already. Yet recently it has re-emerged in modern studies of the universe, albeit in a slightly different guise. Modern cosmologists are not only interested in the structure and history of our universe but also in the other types of universe that might have been. Our universe has many special and (to us at least) surprising properties that we want to evaluate in order to see if they could have been otherwise. This means that we have to be able to produce examples of 'other' universes so as to carry out comparisons.

This is what modern cosmology is all about. It is not just an exercise in describing our universe as completely and as accurately as possible. It seeks to place that description in a wider context of possibilities than the actual. It asks 'why' our universe has some properties and not others. Of course, we might ultimately discover that there is no other possible universe (whose structure, contents, laws, age and so forth are different in a way we can conceive of) apart from the one we see. For a long time, cosmologists were rather expecting – even hoping – that would turn out to be the case. But recently the tide has been flowing in the opposite direction and we seem to be faced with many different possible universes, all consistent with Nature's laws. And, to cap it all, these other universes may not be only possibilities:

they may be existing in every sense that we attribute to ordinary things like you and me, here and now.

THE IMPORTANCE OF PLACE

> And [Jacob] dreamed, and behold a ladder set up on the earth,
> and the top of it reached to heaven: and behold the angels of
> God ascending and descending on it.
>   Book of Genesis[6]

People have been talking about the universe for thousands of years. It was *their* universe, of course, not to be confused with ours. For many it would have been just the land as far as they could journey. Or maybe it was the night sky of planets and stars that could be seen with the naked eye. Most ancient cultures tried to create a picture or a story about what they saw around them, whether it be in the sky, on the ground or under the sea.[7] This drawing of a bigger picture was not originally driven by an interest in cosmology, but was simply important to convince themselves, and others, that things had a meaning and *they* were part of that meaning. To admit that there were parts of reality about which they had no conception or control would have created a dangerous uncertainty. This is why ancient myths about the nature of the universe always seem so complete: everything has a place and there is a place for everything. There are no 'maybes', no caveats, no uncertainties and no possibilities awaiting further investigation. They really were 'theories of everything', but they are not to be confused with science.

Your time and place on Earth influenced the sense you would make of the universe around you. If you lived near the Equator then the apparent motions of the stars each night were clear and simple. They rose, passed up and over your head throughout the night and then descended to set on the opposite horizon. Every night was the same and it felt as though you were at the centre of these celestial movements. But if you lived far from the Tropics the heavens looked very different. Some stars rose above the horizon and set later that night, they came straight up and over your head before falling back to the horizon. Others never rose or set and were always above the horizon.

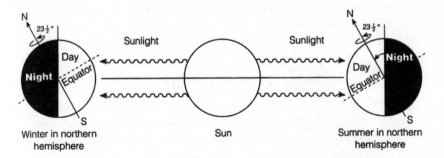

Figure 1.1 The Earth's rotation axis, running through the North and South Poles, is tilted at about 23.5 degrees with respect to the vertical perpendicular to the Earth's orbital plane.

They seemed to trace out circles around a great centre in the sky, like they were pinned to a wheel turning on its axis. It must have made you wonder what was special about that place around which the stars turned. Many myths and legends about the great millstone in the sky were framed by the inhabitants of northern latitudes to make some sense of that great nightly swirl of stars.

The reason for this variation in the appearance of the night sky around the world is a tilt of the axis around which the Earth rotates each day (Figure 1.1). As the Earth orbits the Sun, the line through the Earth's North and South Poles[8] around which it rotates each day is not perpendicular to the line its orbit traces. It is tilted away from the vertical at about 23.5 degrees. This has many remarkable consequences: it is the reason for the seasons. If there were no tilt then there would be no seasonal changes in climate; if the tilt were much larger then the seasonal variations would be far more dramatic. However, if you know nothing about the motion of the Earth around the Sun, or the tilt of its axis, and merely look at the stars in the sky each night, the tilt ensures that there will be a very different sky on view at different latitudes on Earth.

If we extend the line from the South to the North Pole out into space it points in a direction that we call the North Celestial Pole and away from the South Celestial Pole. As the Earth rotates, at night we will see the fixed stars apparently rotating past in the opposite direction across the sky. If they remain visible they will be completing a great circle on the sky each time the Earth completes a daily revolution.

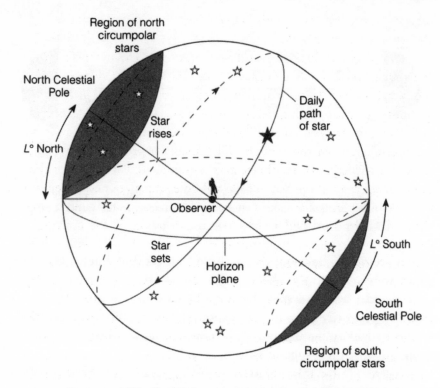

Figure 1.2 The celestial sky seen by astronomers who are located at a latitude of L degrees north. Only half of the sky is visible to them at any moment. Some stars, the north circumpolar stars, are so close to the North Celestial Pole that they never set below the horizon. A second group, around the South Celestial Pole, called the south circumpolar stars, are never seen by the same astronomers because they do not rise above the horizon.

However, not all of these circular paths across the sky will be completely visible to us because part of the path will lie below our horizon. In Figure 1.2 we show what a sky watcher living at a latitude of L degrees in the northern hemisphere will see on a clear night.[9]

Our sky watcher's horizon divides the sky in half. Only the part of it above the horizon can be seen at any moment. Observing from a latitude of L degrees north means that the North Celestial Pole lies L degrees above the horizon and the South Celestial Pole lies L degrees below it. The Earth's rotation makes the sky appear to rotate in a westerly direction around the North Celestial Pole. Stars are seen to rise on the easterly horizon and then move up the sky before reaching their highest point, or 'zenith', after which they descend and set on the westerly horizon.[10]

There are two groups of stars that are not seen to follow this nightly rise and fall in the sky. Stars inside a circle that extends L degrees from the North Celestial Pole complete their apparent circles in the sky without ever disappearing below the horizon. If the sky is clear and dark they can always be seen.[11] For European sky watchers today, they include the stars in the Plough and Cassiopeia groups. Conversely, there is a collection of southern stars within a circular region of the same extent around the South Celestial Pole which are never seen by the southern-hemisphere sky watcher in our picture. They never rise above his horizon.[12] This is why the constellation of the Southern Cross can never be seen from northern Europe. Crucially, we can see that the size of these regions of the sky that are always visible or invisible varies with the latitude of the sky watcher. As your latitude increases and you move away from the Equator, so the sizes of these regions increase as well. In Figure 1.3 we show how the sky would appear to sky watchers at three very different terrestrial latitudes.

At the Equator, the latitude is zero and there are no regions of ever-visible or never-visible stars. An equatorial sky watcher can glimpse every bright star, although the two Celestial Poles are lost in the haze down on the far horizons in practice. The stars rise and ascend to their highest points in the sky. As each star rises, its direction remains relatively constant and is an excellent navigational beacon for wayfaring on land or sea throughout the night. There is almost no sideways motion in the darkness and the sky seems to be very symmetrical and simple. Our sky watchers gain the impression that they are at the centre of things, beneath a celestial canopy of over-arching and predictable motions that seem to be there for their convenience. The universe looks as if they are central to it.

At the extreme case of the North Pole, the latitude is 90 degrees and the visible stars neither rise nor set. They move in circles around the sky. The Celestial North Pole is directly overhead and all the stars circle around it. It looks like the focal point of the universe and we are directly beneath it.

At more temperate northerly latitudes, like that of ancient Stonehenge in Britain at 51 degrees, there is an in-between situation. Stars lying within 51 degrees of the Celestial Pole will be seen to

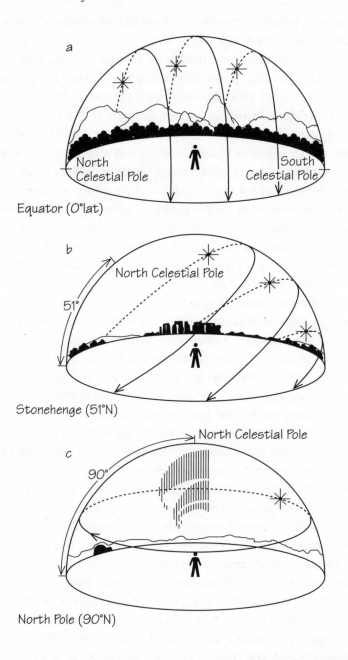

a

North
Celestial Pole

South
Celestial Pole

Equator (0°lat)

b

North Celestial Pole

51°

Stonehenge (51°N)

North Celestial Pole

c

90°

North Pole (90°N)

Figure 1.3 The appearance of the night sky seen from three different latitudes on Earth. It differs because of the change in position of the Celestial Pole around which the stars appear to rotate: (a) at the Equator; (b) at the latitude of Stonehenge in England; (c) at the North Pole.

complete concentric circles on the sky with the Pole as their centre. Other stars will rise above the horizon, ascend to their zeniths, then descend and set. The sky appears extremely lopsided. Different stars follow different paths between their rising and setting. Most striking of all, though, is the great swirl of stars in the direction of the Celestial Pole, all circling it as if it is the hub of a great cosmic wheel (Figure 1.4). For those sky watchers who know nothing of astronomy, or the motion of the Earth, there seems to be a special place in the sky.

This is one reason why there is a geographical dimension in the myths about the sky and the nature of the universe around us. Far from the Equator, in Scandinavia and Siberia, we find legends of the great circle in the sky: the millstone at whose centre the gods reside. The nearest star to the centre of the celestial swirl was given a special importance, hosting the throne of the sovereign of the universe around whom all the stars were arrayed.[13]

We will not be interested in tracing these myths any further here. We simply want to highlight how difficult it was to come up with a picture of the universe from an earthbound vantage point. There are significant biases that you will be unaware of when you know nothing about the stars and the rotation and orientation of the Earth.

Even when sophisticated early civilisations started making astronomical observations they still encountered the influence of our particular vantage point. We are confined to a small planet which, along with many other planets, orbits a star. Today we know about this solar system of planets and the hundreds of distant stars which have been found to have other planets around them (more than 500 at present). This familiarity makes it easy for us to forget how difficult it was to get away from an Earth-centred view and understand the motions of the other planets. As a very simple example of the difficulty, let's think about our view from Earth of the motion of a planet like Mars. We will assume that both the Earth and Mars orbit around the Sun in circular orbits, and that the radius of the orbit of Mars is about one and a half times larger than that of the Earth's orbit (Figure 1.5a). Earth takes one year to complete its orbit and we shall assume that Mars takes twice as long to complete its orbit. Now work out the difference between the two orbits as time passes. This tells us the apparent motion of Mars as seen from Earth. A graph of this is shown in Figure 1.5b.

Figure 1.4 A long-time exposure in the direction of the North Celestial Pole records the circular star trails around the Pole, which is located just above the top of the tree in the centre.

This curious heart-shaped loop with a cross-over (called a 'limaçon') is interesting. As we go from the top right towards the left, we see that Mars is moving away from the Earth. When it drops down to cut the horizontal axis at the point -5 the two planets are on opposite sides of the Sun and as far away from each other as possible. Then as Mars starts to return towards the Earth something very strange happens. Mars approaches the Earth and looks as if it is going to collide with it. But then it reverses its direction and moves away again, to resume its long period of motion away from the Earth. This 'retrograde' motion of Mars can be detected with the naked eye over a period of several nights during a period of close approach. We see that it arises each time the two planets are moving towards their distance of closest approach. If we look instead at one of the distant outer planets, like Saturn, whose orbit takes 29.5 Earth years to complete, there will be several occasions when the Earth–Saturn relative motion is in the vicinity of closest approach during each complete orbit of Saturn and there will be several loops in the picture of the apparent orbit.[14]

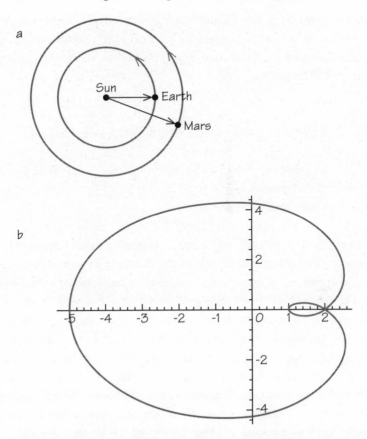

Figure 1.5 The apparent motion of the planet Mars observed from Earth. (a) The orbits of Earth and Mars, which are assumed to be circular, with the radius of Mars's orbit approximately 1.5 times that of Earth's. Mars takes about two years (687 Earth days) to complete its orbit. (b) The two-year orbit of Mars viewed from the Earth, traces out this looped, heart-shaped figure, called a 'limaçon'. Mars moves away at first, reaching its maximum separation of –5 when the Earth and Mars are on opposite sides of the Sun. Mars then returns to its closest approach to Earth but suddenly reverses direction and moves away. Then it changes direction on the sky and starts to move away again.

The lesson we learn from this is that motions in the sky are very difficult to interpret if you do not have an overall picture – or theory – of the motions. An early astronomer watching Mars over two years would have seen it moving away from us, then towards us, before being apparently repelled and sent away again. What forces might be acting? Why does the motion change direction? These are difficult questions to answer if you are located on Earth, unaware that all the orbits (including your own vantage point) are revolving around the Sun at different rates.

ARISTOTLE'S SPHERICAL UNIVERSE

An expert is a person who avoids the small error as he sweeps on to the grand fallacy.
  Benjamin Stolberg

A complicated picture of these apparent celestial motions arose because of a philosophical view of the universe introduced by Aristotle in about 350 BC in an attempt to simplify matters. Aristotle believed that the world did not come into being at some time in the past; it had always existed and it would always exist, unchanged in essence for ever. He placed a high premium upon symmetry and believed that the sphere was the most perfect of all shapes. Hence, the universe must be spherical. In order to accommodate the objects seen in the sky, and their motions, Aristotle proposed a complicated onion-skin structure containing no fewer than fifty-five nested spheres of transparent crystal, centred on the Earth, which is also assumed to be spherical in shape (an assumption that is very hard to reconcile with what he could see!). Each of the observed heavenly bodies was attached to one of these crystal spheres, which rotated at different constant angular speeds. Various extra spheres existed between those carrying the planetary motions. In this way, Aristotle could both explain observations and predict new things that might be seen. It had many features of a modern scientific theory – and many that are unrecognisable. The outer sphere of the stars in Aristotle's picture was a realm where material things could not exist – a spiritual realm. All the motions we see were initiated by a Prime Mover acting at the boundary of this realm and causing the outer sphere to rotate. The rotation was then communicated inwards, sphere by sphere, until the whole cosmos was in perfect rotational motion. By tinkering with the speeds at which the different spheres rotated, many of the features of the night sky could be explained.

a

b

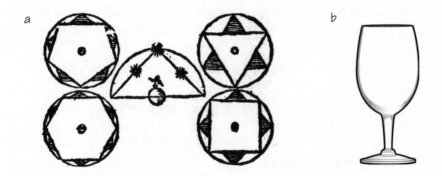

Figure 1.6 (a) A rotating sphere always occupies the same volume of space but other polyhedral shapes create a 'void' when they rotate. This Aristotelian 'proof' of the sphericity of the Earth is shown in Robert Recorde's picture taken from his book *Castle of Knowledge* (1556). However, a wine-glass-shaped universe rotating about its vertical axis, shown in (b), also satisfies Aristotle's requirement that its rotation leave and create no void region.

Aristotle's philosophy was later absorbed and remoulded by medieval Christian thinkers who identified the Prime Mover with the God of the Old Testament and the outermost sphere with the Christian heaven. The centrality of the Earth was concordant with the central part played by humanity in the medieval world picture.

An important feature of the spherical shape for the Earth and all the other outer spheres was the fact that when a sphere rotates it does not cut into empty space where there is no matter and it leaves no empty space behind – see Figure 1.6a, which shows a sixteenth-century description by the eminent Tudor mathematician and physician, Robert Recorde (1510–58). A vacuum was impossible. It could no more exist than could an infinite physical quantity.[15] A stationary spherical Earth always occupies the same portion of space as it rotates. If it were a cube this would no longer be true.[16] In fact, in Aristotle's argument the sphere isn't the only possible shape a rotating Earth could have if it is not to leave or enter a void region of space. A wine glass will do just as well.[17]

Aristotle did not think of motion as being created by forces between objects in the way that we (following Newton) think of gravity. Forces were innate properties of the objects themselves; they moved in a manner that was 'natural' for them. Circular motion was the most perfect and natural movement of all.

PTOLEMY'S 'HEATH ROBINSON' UNIVERSE

> I used to be an astronomer but I got stuck on the day shift.
> Brian Malow[18]

We have already seen that in a solar system in which the Sun lies at the centre and all the planets orbit around it at different speeds you will see strange movements on the sky; other planets will seem to move backwards for a short period. This is an illusion created by our motion relative to those planets. We are all orbiting at different angular speeds and so we sometimes see other planets exhibit unusual counter-motions on the sky. Aristotle and his followers needed to explain those observations.

A solution to this challenging problem was first found by Claudius Ptolemy in about AD 130. It was the nearest thing to a 'Theory of Everything' in the ancient world and it lasted for more than 1000 years. Ptolemy's task was to reconcile the complicated motions of the planets, and all their retrograde movements, with Aristotle's rigid specifications that the Earth was at the centre of the universe. All the other bodies moved in uniform circular orbits at different constant angular speeds around the Earth, and no bodies in the universe can change their brightnesses or other intrinsic properties (Figure 1.7). This was quite a challenge.

Ptolemy addressed this huge problem in his book *The Almagest* ('The Greatest') by considering the circular orbit of a planet, or the Sun, around the Earth to trace out the motion of a point (or 'deferent' as it was termed). This point in its turn then acted as the centre of another smaller circular motion of that planet, called an 'epicycle', along which the planet moved.[19] The overall motion of the planet looks like a circle with a continuous corkscrewing wiggle, shown in Figure 1.8

The overall motion of a planet like Mars relative to the Earth would be a circular orbit around a point that was itself orbiting the Sun in a circle. Ptolemy could have elaborated this further by adding further

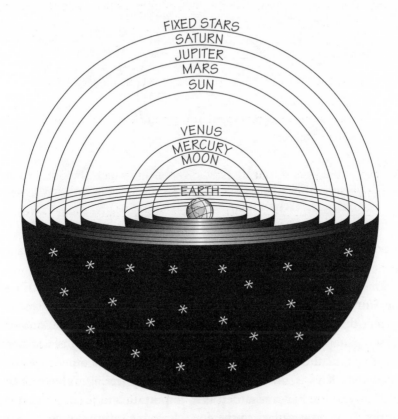

Figure 1.7 The universe model of Aristotle and Ptolemy.

epicycles (circles moving in circular orbits) to the orbit of that planet around the Sun. More and more of these epicycles were added by his medieval successors as they sought for ever greater accuracy.[20]

There were a very large number of things that could be changed in order to make all the features of the moving planets, and the Sun, match observations very accurately. The retrograde motion seen from Earth is well described by the addition of the epicycles. For half of the planet's orbit around its small epicycle it moves in the same sense as the motion around the deferent; but for the second half of the orbit around the epicycle it is moving in the opposite direction and we will observe a retrograde motion. The planet, as seen from Earth, would occasionally slow down, stop on the sky, and then reverse, before slowing down, stopping and then reverting to move in the

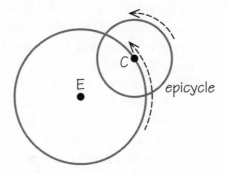

Figure 1.8 Epicycles. A planet, P, moves in a small circle, the epicycle, whose centre, C, follows a larger circular orbit.

opposite direction. This is a real retrograde motion and not an illusory one created by our different orbital speeds around the Sun.

This early response to the complicated motions of the planets and the Sun relative to the Earth shows how difficult it is to arrive at a correct description of the universe from observation alone, or from a very general philosophical principle. If the Aristotelians had been more critical they would have had to grapple with other awkward problems. Why is the Earth not perfectly spherical? Why was the centrality of the Earth regarded as so important and yet other circular motions could take place in the epicycles which were not centred on the Earth? Why was the idea of displacing the centre of each planet's deferent circle away from the Earth accepted? The displacement may have been small but the Earth is either at the centre of the universe or it isn't.

COPERNICAN REVOLUTIONS

> If the Lord Almighty had consulted me before embarking on
> creation thus, I should have recommended something simpler.
>   Alphonso X of Castile[21]

Ptolemy's model of the universe, with the Earth at its centre, was an intricate human conception. It wasn't right but it had so many ways of being tweaked to handle awkward new observations about planetary movements that it survived, largely unchallenged, until the fifteenth century. This elastic feature even led to the word 'epicycles' becoming a pejorative term to describe any slippery or overcomplicated scientific theory. If you have to keep adding new details into the workings of a theory to explain every new fact that comes along then your theory has little explanatory power. It is as if you have a new theory about cars which predicts that all cars are red. On Monday morning you step outside and see a black car, so you modify your theory to predict that all cars are red, except on Mondays when some are black. Lots of black and red cars pass by. All seems well. But then in the afternoon a green car drives by. Okay, all cars are red and black on Mondays except after noon, when some cars are green. You can see the way it's going. This is a theory of cars that has a series of correcting 'epicycles'. Every new fact is accommodated by a little modification so as to maintain the grand assumption you started out with. At some stage you should get the message and start again.

This is of course an exaggerated example. Ptolemy's theory was more sophisticated. Each time an epicycle was added it introduced a new *smaller* correction to accommodate a finer detail of the observed motions. This theory was one of the first examples of a convergent approximation process in action. Each addition to the model is smaller than the last and produces a better description of the observations.[22] It worked rather well for most purposes despite having the wrong overall picture of the solar system and the wrong celestial body (the Earth instead of the Sun) at its centre! It took a very persuasive case to turn the tide of opinion against it.

Nicolaus Copernicus is generally regarded as a revolutionary – the scientist who dethroned humanity from its central position in the universe. The reality has turned out to be more complicated and far less dramatic, and if he was a revolutionary at all, he was certainly a reluctant one.[23] Copernicus's great book, *De revolutionibus orbium coelestium* ('On the Revolutions of the Celestial Spheres'), was delivered to the printers in 1543, shortly before his death, and its impact was muted. Not many copies were printed and few of those were ever read. Yet, in time, Copernicus's perspective became the rallying point for the transformation of our view of the universe. It would eventually displace the ancient Ptolemaic picture of a planetary system with the Earth at its centre in favour of a Sun-centred model that remains with us today.[24]

The advances in printing during the early sixteenth century meant that Copernicus's book could be printed with diagrams embedded in the text at the points where they are discussed. The most famous of his diagrams (see Figure 1.9) shows a simple model of the solar system with the Sun located at its centre. The outermost circle marks the boundary of the 'immobile sphere of the fixed stars' beyond our solar system. Each of the other six circles marks a sphere of motion for the orbits of the six, then-known, planets. Going from the outside inwards, they denote the planets Saturn, Jupiter, Mars, Earth (with its adjacent crescent Moon), Venus and Mercury, respectively. All follow circular tracks around the central Sun (*Sol*). The Moon was believed to move in a circle around the Earth.

The systems of Copernicus and Ptolemy were not the only pictures of the Sun and the planets that were on offer in the sixteenth and seventeenth centuries. Figure 1.10, taken from Giovanni Riccioli's *Almagestum Novum* ('The New Almagest')[25] of 1651, nicely summarises the world pictures on offer to astronomers in the post-Copernican era. It shows six different models of the solar system (labelled I–VI).

Model I is the Ptolemaic system with the Earth located at the centre and the Sun's orbit around it lies outside the orbits of Mercury and Venus.

Model II is the Platonic system, where the Earth is also central. The Sun and all the planets are in orbit around it but the Sun lies inside the orbits of Mercury and Venus.

Model III is the so called Egyptian system, in which Mercury and Venus revolve around the Sun, which, along with the outer planets, revolves around the Earth.

Model IV is the Tychonic system of the great Danish astronomer Tycho Brahe (1546–1601), in which the Earth is fixed at the centre and the Moon and the Sun revolve around the Earth but all the other planets revolve around the Sun. The orbits of Mercury and Venus are therefore partly between the Earth and the Sun, while the orbits of Mars, Jupiter and Saturn encircle both the Earth and the Sun.

Model V, called the semi-Tychonic system, was invented by Giovanni Riccioli himself. In his model the planets Mars, Venus and Mercury orbit the Sun, which, together with Jupiter and Saturn, orbits the Earth. Riccioli wanted to distinguish Jupiter and Saturn from Mercury,

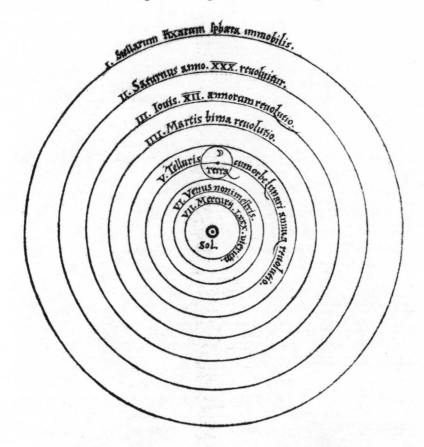

Figure 1.9 Copernicus's heliocentric picture of the solar system, published in 1543. The diagram is labelled in Latin and shows concentric spheres around the central Sun. The fixed outer sphere of the stars (I) surrounds rotating spheres (II–VII) containing the orbits of Saturn, Jupiter, Mars, Earth (with its Moon marked as a crescent lune), Venus and Mercury.

Venus and Mars because they were known to have moons like the Earth (the two moons of Mars had not yet been discovered) and so their orbits must be centred on the Earth rather than on the Sun.

Model VI is the Copernican system we have just seen illustrated in Figure 1.9.

This selection of ancient astronomical views of the universe has taught us some simple lessons. It is not easy to understand the universe just by looking at it. We are confined to the surface of a particular type of planet in orbit, with others, around a middle-aged star. As a

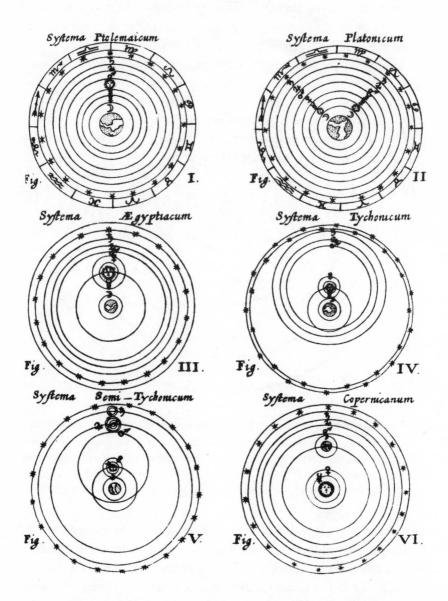

Figure 1.10 The six major world systems of 1651 shown by Giovanni Riccioli in his book *The New Almagest*.

result, what we see in the night sky is strongly determined by *where we are located* on the Earth's surface, *when we look*, and by any preconceptions we might have about where *we should be* in the grand scheme of things. Our world view predetermines our world model.

As our view of the universe expanded so these problems got bigger too. In order to make progress we need to be able to describe and predict the celestial movements we see in our part of the universe. But, ultimately, we want to know what the whole universe is like. The first decisive steps in that direction were taken by astronomers in the eighteenth century. Let's follow them next.

2 The Earnestness of Being Important

Symmetry calms me down. Lack of symmetry makes me crazy.
  Yves Saint Laurent

SPECIAL TIMES AND SPECIAL PLACES

There isn't a Parallel of Latitude but thinks it would have been
the Equator if it had had its rights.
  Mark Twain

Nicolaus Copernicus has given his name to an entire philosophy of
the world. In science an 'anti-Copernican' view is a pejorative description of thinking that presumes the centrality of humanity. In astronomy,
the Copernican 'Principle' is often invoked to add prestige to the idea
that we should always assume our position in the universe is not
specially privileged. Instead of thinking the Earth is the centre of the
universe, like the ancients, we should assume that the universe is much
the same everywhere and build our theories accordingly. And so the
Earth is expected to be a typical planet orbiting a typical star in a
typical galaxy in the universe.

While the removal of the Earth and humanity from the centre of
the universe is an important general lesson for scientists, we have
come to appreciate that it contains pitfalls of its own if over-zealously
pursued. For while we have no reason to expect that our position in
the universe is special in *every* way, we would be equally misled were
we to assume that it could not be special in *any* way.

We now understand that life can only exist in regions of the universe that have certain features: obviously we could not exist at the centre of a star where no atoms could survive, or in a region of the universe where the density of matter is too low for stars to form.[1] If 'typical' places in the universe are ones that have an environment that does not allow life to develop and persist then we cannot be in a typical position. This simple moderation of the Copernican perspective plays a crucial role in the testing of predictions in modern cosmology.[2]

Location is not, as the estate agents say, everything. We must also consider our place in history. If the universe changes its overall properties with time, say, getting hotter or cooler as it ages, then we may find that stars and planets and life can only be supported during particular intervals of cosmic history. This type of bias is linked to many of the most significant features of the expanding universe we observe today. The universe appears very old because the building blocks of chemical complexity, nuclei of elements like carbon, nitrogen and oxygen, are made in the stars, by a slow process of nuclear burning that culminates in a supernova explosion which spreads those life-supporting elements into space. Eventually, those ingredients find their way into planets, and into you and me. This process of stellar alchemy takes billions of years to run its course. So we should not be surprised to find that our universe is so old. We could not exist in one that was significantly younger: it would not have had time to generate the building blocks needed for complexity of life.

In the future there will be a time when the last star exhausts its nuclear fuel and 'dies', collapsing into a dense endlessly cooling remnant, or a black hole. Perhaps this means there is a time after which no life can survive in the universe. Some people find this a very unsatisfactory state of affairs and believe that life will never die out.[3] Certainly, life as we know it – carbon-based and biochemical – cannot survive indefinitely. But if we look at the direction in which our advanced technologies are evolving, there is hope. Continual miniaturisation allows resources to be conserved, efficiency to be increased, pollution to be reduced, and the remarkable flexibilities of the quantum world to be tapped. Very advanced civilisations elsewhere in the universe may have been forced to follow the same technological path. Their nano-scale space probes, their atomic-scale machines and nano-computers, would be imperceptible to our coarse-grained surveys of the universe. Their waste energy output would be small, leaving little trace. This may be the low-impact evolutionary path you need to follow in order to survive into the far, far future.

DEMOCRATIC LAWS

One law for all
Roman law

After Copernicus, his Sun-centric picture of our solar system was gradually refined and eventually described mathematically by a new theory of motion and gravity created by a young man from Lincolnshire, called Isaac Newton (1643–1727). Newton's law of gravitation and three laws of motion dominated the way physicists and engineers understood the world for nearly 250 years. They transformed all previous pictorial descriptions of motion into a precise mathematical one. They provided equations ('laws' of change) whose solutions (the 'outcomes' of those laws) predicted successfully what we should see as time passes, the motions of the Moon, and the planets. One of those predictions was that the orbit of a planet around the Sun will not be circular, as Copernicus had assumed, but elliptical with the Sun located at one of the two focal points of the ellipse (Figure 2.1).

Newton's three laws of motion can be paraphrased as follows:

First law: Bodies acted upon by no forces remain at rest or moving at constant speed in a straight line.

Second law: The rate of change of momentum of a body equals the force applied to it.

Third law: To every force there is an equal and opposite force of reaction.

These laws hide many remarkable insights. The first one refers to bodies acted upon by 'no forces'. For who has ever seen such a body? It is an idealisation that Newton recognises as a fundamental benchmark. In the past, most people thought that bodies acted upon by no forces just slow down and stop. But Newton realised that the slowing is caused by the action of other forces, like friction and air resistance.

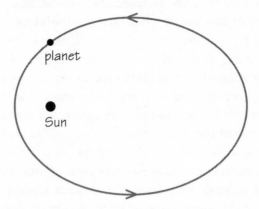

Figure 2.1 An exaggerated elliptical orbit of a planet showing the location of the Sun at a focal point of the ellipse.

He was acutely aware of all the forces at work in a given situation and was able to imagine what would happen if none of them were acting, an entirely idealistic situation.

When Newton talks about bodies moving or being at rest, we could ask: 'At rest relative to what?' In fact, he refers all those motions to an imaginary fixed stage of space marked out by the distant stars which he takes to be unchanging and unmoving, and which became known as 'absolute space'. Newton's laws dictated how things moved around and acted out their parts on that stage. Nothing they did could ever change its fabric.

Newton realised that only special actors on this cosmic stage would see his simple laws to be true. They would need to move so that they did not accelerate or rotate relative to the far distant stars that fixed his unchanging 'absolute' space. Suppose that an astronaut looks out of the window of a rotating spaceship. He will see the distant stars rotating past in the opposite direction to the spin of the spaceship. These stars are rotating in circles and relative to him they are accelerating, but they are not being acted upon by any forces. So, for the spinning astronaut, Newton's first law does not hold and the form of the second law of motion he deduced would become more complicated.[4]

Newton's formulations of his laws of motion highlight the existence of a form of Copernican principle for the laws of Nature as well as for their outcomes. They required special observers, for whom the laws of motion look simpler than they do for everyone else. Yet surely the true laws of Nature, rightly expressed, should look the same for all observers, regardless of their motion or their position. No one should be privileged to find them simpler than everyone else.

Armed with Newton's laws, physicists and astronomers could try to make sense of all the motions they saw in the heavens. They could try to understand the distributions of stars and how things got to be as they are today from simpler beginnings. They lacked the telescopic power that we possess today so their view of the cosmic landscape was limited in scope. Yet, gradually pictures were constructed that accounted for the distribution of stars and linked these astronomical pictures to what we know about physics and motion. And, crucially, they began to think about what Newton's laws of motion might tell us about how the universe changes.

THE CHANGING UNIVERSE

Round like a circle in a spiral
Like a wheel within a wheel
    Alan and Marilyn Bergman

In the centuries that followed Newton, our appreciation of the scope and scale of the universe grew steadily. Thomas Wright (1711–86), a clockmaker, self-taught astronomer, surveyor and architect from Durham in the north of England, was the first to seek a detailed picture of the Milky Way, the band of stars, gas, dust and light that had been known and admired by all who had studied the sky from ancient times.[5] He recognised that the evidence provided by the early telescopes was showing that the sky was not randomly sprinkled with stars. Rather, they displayed very distinctive clustered patterns which we were a part of, looking outwards from inside of one of those clusters. What was the real three-dimensional pattern that gave the Milky Way its observed appearance?

Wright proposed two possibilities. The first imagined a clustering of stars in a disc of flat rings, rather like those we are familiar with around Saturn, circling around the centre of the Milky Way. The centre

a

b

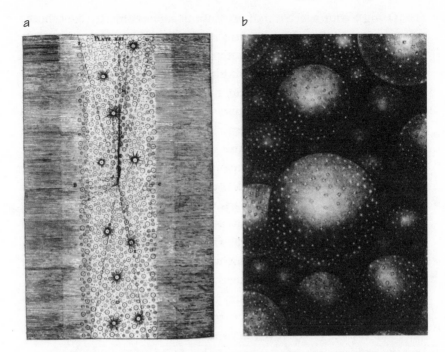

Figure 2.2 (a) Thomas Wright's model of the Milky Way galaxy in which the stars are uniformly distributed in a disc-shaped slab of space. (b) Wright's never-ending universe containing an infinite number of galaxies that look like bubbles in infinite space. Both illustrations are from his 1750 book, *An Original Theory of the Universe.*

was the 'centre of creation' from which 'all the laws of nature have their origin'. The second possibility was that the stars were clustered as if on the surface of a sphere. The Milky Way was a slice through the edge of this shell, reflecting the fact that we did not live near the centre of the galaxy (see Figure 2.2a).

Wright's imagination flew higher still. He could see no reason why there should be only one of these great collections of stars. He imagined an endless number of them all over the universe, each a centre of created stars, some spherical, some disc-like. The faint images all over the night sky suggested to him that they might all be Milky Ways, creating an 'endless immensity . . . not unlike the known universe', which is shown in Figure 2.2b.

Wright laid great stress on poetic and pictorial descriptions of the universe. He drew a large (3 x 2 square metres) plan of the universe illustrating a wide spectrum of astronomical phenomena, like eclipses and cometary orbits. He was also inspired to think of an unending universe of solar systems, each with its own system of planets orbiting around its central star, by John Milton's mention of suns and other worlds in *Paradise Lost*. For Wright, our Sun led to other suns, and the Earth must be just one of many planets. He estimated that there would be more than 3,888,000 stars in the Milky Way, and '60,000,000 planetary worlds like ours', yet this is but a tiny portion of the whole night sky.

Wright's speculations and model-building were an important extrapolation of the Copernican emphasis on our solar system, in order to come to terms with a bigger universe. The idea that the universe was populated by huge numbers of galaxies (or 'island universes'), of which our Milky Way is but one, did not win acceptance until 1921, after the American astronomer Heber Curtis famously argued against Harlow Shapley in a great public debate at the Smithsonian Institution in Washington DC that the spiral nebulae seen on the night sky were indeed far-distant galaxies like our own. Shapley argued, unconvincingly to astronomers at that time, that the Milky Way was the entire universe.

Ironically, Wright's far-sighted work is mainly remembered because of the ways in which it was taken up by other scientists, who were better equipped to add further substance to it. Wright didn't make any useful further contribution to observational astronomy as he moved on to follow an architectural career,[6] but one of his most sharp-minded young readers was fascinated by the picture of the universe he created. In 1751, at the age of twenty-seven, Immanuel Kant read a (not entirely reliable) second-hand account of Wright's work in a Hamburg newspaper. Four years later, in response, he wrote an anonymous account of the universe, entitled *Universal Natural History and Theory of the Heavens*, which enjoyed only a very limited circulation because the publishers went bankrupt and the printed copies ended up being seized by the bailiffs. It came to prominence only a century later, when it was drawn to astronomers' attention by Hermann von Helmholtz in a public lecture in Germany.[7]

Kant enthusiastically took up Wright's picture of the Milky Way, suggesting that his rings of stars were a rotating disc of stars in which the inward attraction of gravity was counter-balanced by the outward

Figure 2.3 Immanuel Kant (1724–1804).

centrifugal force of reaction to their rotation around the centre of the galaxy. Without gravity the disc of stars would disperse; without rotation all the stars would collapse in upon themselves. All the nebulae[8] that astronomers could see with their telescopes, Kant argued, were also just spinning discs of stars.

This similarity throughout the universe was a reflection of the universality of Newton's laws of gravity and motion. The systems of stars varied only in apparent brightness, a reflection of their different distances away from us; their different patterns on the sky were explained by the different orientations they had with respect to our line of sight – like looking at a rugby ball from different angles. Then, Kant went a little further, looking for further patterns that might conceivably exist. If stars were clustered into galaxies like the Milky Way, maybe these galaxies are then gathered into great clusters of galaxies, which in turn are gathered into clusters of clusters, and so on for ever. The scheme doesn't seem entirely consistent as it would require the galaxies to rotate around their cluster centres to form discs and these clusters to form great rotating discs as well, but it was a brilliant attempt to use Newton's laws to understand structures in the universe beyond the solar system.[9]

The most striking feature of Kant's picture of the universe was that it was evolutionary: it was a cosmos that changed with time as stars came and went.[10] His universe was infinite in extent, so it didn't have a true centre. But there could be special places, picked out by the fact that the density there was highest: our solar system is located at such a place. Life and organisation were seen propagating outwards from this centre like a spherical wave, leaving new worlds in their wake. On each scale where new structures formed, they were maintained in equilibrium by a balance between the force of gravity pulling them towards the centre and a centrifugal force of rotation pushing outwards, just as in the Milky Way.[11] While the leading edge of this wave was full of other solar systems, there was a trailing edge populated by dead worlds that had exhausted their resources and fallen into decay (Figure 2.4). The evolution of new and fruitful structures takes place in the outer part of the expanding shell. This is the creative frontier of the universe of matter. The old decayed material in the central region is a residue of the first structures to form. Yet it is not only a cosmic graveyard. This material could rearrange itself and be recycled so as to create the ordered stars and planetary systems of the future, like a 'Phoenix of nature', and so the universe would continue for ever: the 'creation is never finished or complete. It has indeed once begun, but it will never cease' and 'worlds and systems perish and are swallowed up in the abyss of eternity; but at the same time creation is always busy constructing new formations in the heavens, and advantageously making up for the loss.'[12]

Kant saw all this as part of a grand Divine plan in which God engages in incessant creation of new and greater worlds. All known life has a finite span and we cannot be expected to be exempt from this law. Yet Nature is inexhaustibly rich, allowing 'whole worlds and systems to quit the stage of the universe, after they have played their parts' as all possible combinations of matter are gradually explored.

Kant identifies what he calls 'a certain law' which produces cycles in which the oldest structures decay first, while new structures form. Hence 'the developed world is bounded in the middle between the

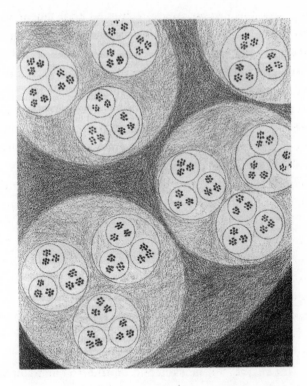

Figure 2.4 Kant's evolving universe consisted of an infinite number of expanding shells, like this, in which the leading edge of each shell produces stars which then gradually fade.

ruins of the nature that has been destroyed and the chaos of the nature that is still unformed.'

Later in his career, Kant would move from his pursuit of astronomy and Newtonian physics to the critical philosophy of knowledge that made him one of the world's most famous philosophers. In his *Critique of Pure Reason* of 1787 he introduced important distinctions between reality and perceived reality. We have to distinguish the objective truth from the truth that our minds, with their particular categories of thought, can comprehend. He compared the impact of his philosophical argu-ments to the Copernican revolution that had occurred in astronomy:

Until now people have assumed that all our knowledge must orient itself to objects . . . [But] now we would like to try to improve our progress in the projects of metaphysics by assuming that objects must be oriented to our knowledge . . . We thus proceed exactly along the main line of thought of Copernicus. When he couldn't make any progress explaining the movements of the heavens by assuming that the whole starry host revolved around the observer, he pinned his hopes for success on a reversal of this assumption, thus making the observer revolve and letting the stars rest in peace. Here we can attempt the same sort of procedure in metaphysics as regards the intuition of objects.[13]

For Kant, the absolute truth about things was unknowable. Only part of it, conditioned by our categories of thought, could be apprehended.[14]

THE NEBULAR HYPOTHESIS

> It is always dark. Light only hides the darkness.
>   Daniel K. McKiernan

Kant also made a contribution to the main rival cosmic theory. In his *Theory of the Heavens* of 1755 he sketched another scenario in which the solar system formed out of a spinning cloud of gas and debris. This idea was developed in a much more precise form by the French astronomer Pierre Laplace (1749–1827) in his popular book *Exposition du système du monde*, which appeared in 1796. This extremely readable account of his ideas about the nature of the universe had a big impact on French (and later European) intellectual life.

Laplace was a major figure in France, a scientific advisor to Napoleon and a distinguished astronomer, mathematician and physicist, who was eventually made a Marquis by the Emperor. He was also an ardent rationalist who wanted to show that it was possible to explain how planets came into existence without any supernatural intervention. The last chapter of his book explained the origins of the solar system by means of a contracting and rotating cloud of material out of which was formed a collection of planets all orbiting in the same plane around the central Sun and spinning on their axes in the same sense.[15] This picture became known as Laplace's 'nebular hypothesis'. It became popular with astronomers who now believed that every blob of light in the night sky was a forming planetary system. This was very different to Wright's scenario, in which those patches of light were identified with whole galaxies like the Milky Way, each containing billions of stars and planetary systems.

Marquis Laplace's picture became the standard Victorian model of the universe. In 1890, the foremost historian of astronomy of the day, Agnes Clerke, went so far as to maintain that:

No competent thinker. . .can now, it is safe to say, maintain any single nebula to be a star system of coordinate rank with the Milky Way. A practical certainty has been attained that the entire contents, stellar and nebular, of the sphere belong to one mighty aggregation.[16]

The Victorian universe was a great cartwheel of stars that formed the Milky Way. The idea that any of those dots of light on the night sky might be external galaxies in their own right was a concept whose star was temporarily fading.

LIFE IN AN EDWARDIAN UNIVERSE

'Why are we here? Because we're not all there.'
  New Tricks[17]

Alfred Russel Wallace (1823–1913) was a great nineteenth-century scientist who today receives less credit than he deserves for his discovery that living organisms evolve by a process of natural selection. Fortunately for Charles Darwin, who had been thinking about the same idea, and gathering evidence for it independently over a very long period of time, Wallace wrote to him to tell him of his ideas rather than simply publish them in the scientific literature, and Wallace and Darwin's theories of natural selection were announced publicly at the same time. Wallace was interested in physics, astronomy and earth sciences and supported Darwin as a colleague over a long period, supplying him with specimens from distant places for his work. In 1903, under the title *Man's Place in the Universe*, he published a wide-ranging study of the factors that make the Earth a habitable place and of the philosophical conclusions that might be drawn from the state of the universe.[18]

DIAGRAM OF STELLAR UNIVERSE (Plan).

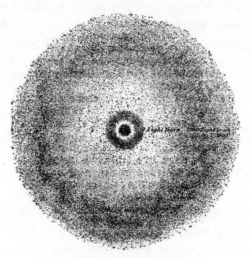

1. Central part of Solar Cluster.
2. Sun's Orbit (Black Spot).
3. Outer limit of Solar Cluster.
4. Milky Way.

Figure 2.5 Lord Kelvin's universe as drawn by Alfred Russel Wallace in 1903, showing the solar system located away from the centre.

Wallace was impressed by a simple cosmological model that Lord Kelvin,[19] the leading British scientist of the day and the President of the Royal Society (1890–95), had developed by using Newton's law of gravitation to explain the fate of huge clouds of material in the universe. Kelvin's interests were extraordinarily broad, and they started early: he was attending lectures at the University of Glasgow from the age of ten and writing important research papers on the structure of the Earth by the time he was fifteen. He developed our understanding of the conservation of energy and the laws of thermodynamics, and introduced the absolute temperature scale, but he was instrumental, too, in the creation and instrumentation of the first transatlantic submarine telegraph cable in 1858. He also found time to design the standard water tap, engineer the heat pump used for central heating and air conditioning, and play an important part in designing the first electric railways.

When he turned his mind to universes, Kelvin was no less incisive. He was able to show that gravity would cause a very large ball of material to implode towards its centre. The only way to avoid falling into the centre was to be in orbit around it, as Kant had proposed. Kelvin's model contained about one billion stars with the same size as the Sun, so that their gravitational pull would create star motions at the speeds observed near us in the universe.[20]

Kelvin's universe, as drawn by Wallace, is shown in Figure 2.5.[21] What is intriguing about his discussion of Kelvin's model of the universe is that he adopts a superficially non-Copernican attitude because he sees how some places in the universe are more conducive to the presence of life than others and that we therefore lie near, but not at, the centre of things.

In Kelvin's cosmological model, material would fall into the central regions where the Milky Way was situated and coalesce with other stars that were already there, generating heat and maintaining the power output over huge periods of time. Wallace's explanation of the vast size of the universe is worth quoting at length:

we have found an adequate explanation of the very long-continued light- and heat-emitting capacity of our sun, and probably of many others in about the same position in the solar cluster. These would at first gradually aggregate into considerable masses from the slowly moving diffuse matter of the central portions of the original universe; but at a later period they would be reinforced by a constant and steady inrush of matter from its outer regions possessing such high velocities as to aid in producing and maintaining the requisite temperature of a sun such as ours, during the long periods demanded for continuous life-development. The enormous extension and mass of the original universe of diffused matter (as postulated by Lord Kelvin) is thus seen to be of the greatest importance as regards this ultimate product of evolution, because without it, the comparatively slow-moving and cool central regions might not have been able to produce and maintain the requisite energy in the form of heat; while the aggregation of by far the larger portion of its matter in the great revolving ring of the galaxy was equally important, in order to prevent the too great and too rapid inflow of matter to those favoured regions . . . For [on] those [planets around stars] whose material evolution has gone on quicker or slower there has not been, or will not be, time enough for the development of life.[22]

Wallace sees the connection between these unusual global features of the universe and the conditions necessary for life to evolve and prosper:

we can dimly see the bearing of all the great features of the stellar universe upon the successful development of life. These are, its vast dimensions; the form it has acquired in the mighty ring of the Milky Way; and our position near to, but not exactly in, its centre.[23]

He also sees that this process of infall and solar power generation from gravitational energy will most likely take a staccato form, with long periods of infall driving the heating of the stars followed by net heat output and subsequent cooling, a period which we have just begun to experience.

Wallace completes his discussion of the cosmic conditions needed for the evolution of life by turning his attention to the geology and history of the Earth. Here, he sees a far more complicated situation than exists in astronomy. He appreciates the host of historical accidents that has marked the evolutionary trail leading to human life and thinks it 'in the highest degree improbable' that the whole collection of features that are conducive to the evolution of life will be found elsewhere. This leads him to speculate that:

such a vast and complex universe as that which we know exists around us, may have been absolutely required . . . in order to produce a world that should be precisely adapted in every detail for the orderly development of life culminating in man.[24]

Wallace was psychologically averse to the idea of a universe populated with other living beings but he believed that the uniformity of the laws of physics and chemistry[25] would ensure that:

organised living beings wherever they may exist in the universe must be fundamentally, and in essential nature, the same also. The outward forms of life, if they exist elsewhere, may vary, almost infinitely, as they do vary on earth . . . We do not say that organic life *could* not exist under altogether diverse conditions from those which we know or can conceive, conditions which may prevail in other universes constructed quite differently from ours, where other substances replace the matter and ether of our universe, and where other laws prevail. But *within* the universe we know, there is not the slightest reason to suppose organic life to be possible, except under the same general conditions and laws which prevail here.[26]

Wallace's approach to cosmology shows how the consideration of the conditions necessary for the evolution of life is not wedded to any particular theory of star formation and development but must be used appropriately in any cosmology we pursue.

THE DECAYING UNIVERSE

> . . . if we consider the case of the whole universe we should be able, supposing we had paper and ink enough, to write down an equation which would enable us to make out the history of the world forward, as far forward as we liked to go; but if we attempted to calculate the history of the world backward, we should come to a point where the equation would begin to talk nonsense – we should come to a state of things which could not have been produced from any previous state of things by any known natural laws.
> William Clifford[27]

During the nineteenth century a new way of looking at the universe – what we now call a paradigm – started to emerge. The Industrial Revolution dominated the Victorian era. Engineering, machines, ships, steam engines and furnaces were driving the economy and scientific developments mirrored these concerns, culminating in the discovery of the laws of thermodynamics.[28] The process of change and progress became an article of faith for philosophers and engineers. Not surprisingly, scientists began to think of the whole universe as a great machine and asked what the laws of thermodynamics had to say about its past and future.

The most far-reaching discovery that physicists had made about heat engines was about their processing of ordered forms of energy – like electrical current or rotary motion – into completely disordered forms, like heat radiation. In 1850, Rudolf Clausius (1822–88) showed that in a closed finite system, from which nothing can escape, this energy processing was a one-way street. Disordered forms of energy, which in 1865 he called 'entropy', could never decrease. This became known as the 'second law' of thermodynamics and is one of the great explanatory principles of science.[29] Yet it is not a law of Nature in the traditional Newtonian sense. It doesn't tell you what happens when a force is applied or when an object falls under gravity: it is a statistical law that governs the behaviour of whole volumes of molecules.

Newton's laws allow all sorts of things to happen that never occur in practice. For example, they allow wine glasses to fall to the ground and smash into fragments (something which we do see) as well as the time-reverse in which many glass fragments come together spontaneously and simultaneously to form a complete wine glass (which we never see). The reason for the difference here is that it is not difficult to contrive the circumstances that lead to the broken glass but it is fantastically improbable to contrive a situation in which all the correctly sized glass fragments start moving at just the right speeds and in the directions needed to create a whole wine glass. Consequently, although Newton's laws allow it, we never see the latter sequence of events in which disorder turns into order: we see the decay of order into disorder instead. It is far more probable.

What if this 'second law' of disorder increase governs the entire universe? In Clausius's words, it means that 'the entropy of the world tends towards a maximum.' In particular, he argued, this ruled out a cyclic universe in which the same general conditions recur or one which dies and rises again like a phoenix from the ashes. It was this question that led to the concept of the 'heat death of the universe'. The one-way slide from order to disorder meant that the universe was apparently doomed to degrade continuously from states of order into states of disorder in the far future.

Eventually, everything would be left in a sea of heat radiation. There would be no stars and planets, no differences in temperature and energy from one thing to another or from place to place. This thermal homogeneity is the death of change and progress, the extinction of that phenomenon we call 'life'. Looking backwards in time, this steady dissolution must have sprung from a past that would have been more ordered. Perhaps the universe even had a beginning in a state of the maximum possible order? Or perhaps the right conclusion to draw would be that the universe cannot be infinitely old or it would have reached a state of complete thermal equilibrium and heat death already?[30]

This concept appealed to those who were still looking to reconcile the idea of a universe that appeared out of 'nothing' a finite time ago with new ideas about evolution and change. Yet the message for the future of humanity was a gloomy one. The revolutionary progress and technological changes that were transforming the industrial world were moving inexorably to an end that was careless of human existence. Suddenly the universe didn't look such a good place to live.

These ideas were taken up by Kelvin in a series of papers and lectures between 1851 and 1854. He was interested in what the second law might tell us about the past as well as the future. Kelvin had a strong religious motivation for deducing a beginning and excluding the conception of an eternal cyclic universe.[31] Yet he was not so happy with the message of heat death, which he didn't accept as an inevitable consequence of the second law. Instead, Kelvin believed the universe was infinite (rather than finite as Clausius's argument required) and thought it was possible that the laws of Nature might change in the future. Others, too, like Ernst Mach, tried to confine the consequences of the second law to individual things, like stars and planets, while denying that the laws of thermodynamics could be applied to the universe as whole. It was not obvious to them that the universe was a closed thermodynamic system, or even that it could be said to be affected by entropy.

The use of the second law to deduce that the universe had a beginning was not confined to Christian apologists. There were also ardent materialists like the logician, philosopher and economist William Jevons (1835–82)[32] who believed that the second law implies that the universe must have had a beginning or there was a time before which the laws of Nature were different. Yet political philosophers like Friedrich Engels, the proponent of dialectical materialism, could only countenance the increase of entropy if the universe was cyclic and treated all arguments about the finiteness of the world or its inevitable dissolution in the heat death as closet arguments for the existence of God, which he rejected absolutely.

The only person who seems to have taken seriously the simple mathematical truth that the ever-increasing nature of the entropy did *not* imply that it had to have been zero a finite time ago[33] was the

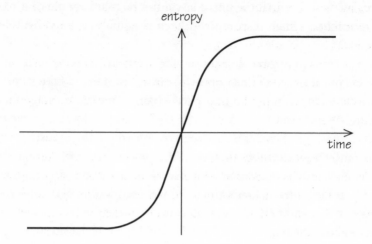

Figure 2.6 A curve that is always increasing but never reached zero in the past and is bounded to the future.

Catholic physicist and historian of science Pierre Duhem (1861–1916). He refused to use it as an argument for the creation of the universe out of nothing in the finite past or the achievement of a total heat death in the future because the continuous increase of the entropy of the universe did not mean it had ever experienced a minimum value, or would ever reach a maximum value in the future. A simple example is shown in Figure 2.6.

The last twist to the thermodynamic perspective was provided in 1895 by Ludwig Boltzmann and Ernst Zermelo, who explored the idea that the universe is infinite and already is in a state of overall thermal equilibrium. None the less, there exist occasional random fluctuations away from this equilibrium state in different places. Some of these fluctuations could be as large as our Milky Way galaxy and provide the sites where life can exist.[34] These large fluctuations are very rare, but so is life.

In fact, the Boltzmann fluctuation argument had been suggested earlier, in 1879, by the English physicist Samuel Tolver Preston. Preston was trained as a telegraph engineer but became something of an expert on thermodynamics and gravity and eventually received his doctorate (at the age of fifty) in 1894 after studying in Germany. Preston was impressed by the vastness of the universe and thought that we couldn't generalise from our little visible patch to draw sweeping conclusions about the whole. He suggested that there are regions in the universe which display properties that are conducive to life but we cannot conclude the whole universe is the same. In particular, we must observe entropy to increase in our part of the universe so that biochemical processes can occur because 'from the fact of our being in existence, we must be in a part which is suited to the conditions of life'. Moreover:

the universe would have the peculiar characteristics of allowing almost indefinite local fluctuations of temperature, of states of aggregation, and of composition, of the matter forming the universe, within regions very extensive . . . the constitution of the vast whole (looked at broadly) remaining uniform throughout.[35]

Preston's theory avoided the horrible conclusion that the laws of physics must have broken down at some time in the past if entropy increased in the same way everywhere.[36] We shall have a little more to say about this idea in chapter 10: it continues to be relevant to cosmology more than 130 years after it was originally proposed.

KARL SCHWARZSCHILD: THE MAN WHO KNEW TOO MUCH

I'm through with counting
The stars above
   The Everly Brothers[37]

During the nineteenth century mathematicians finally woke up to something that had been staring them in the face for centuries. They had failed to take on board the existence of geometrical systems other than Euclid's classic description of lines, points and angles on flat surfaces. The prejudice that regarded Euclid's geometry as the one and only logical system of its type was deeply rooted in beliefs about its correspondence to the universe. It wasn't just a mathematical 'game', a system of starting positions and rules from which you could work out all possible geometrical consequences. It was how the world truly was: a piece of absolute truth about the nature of things. When

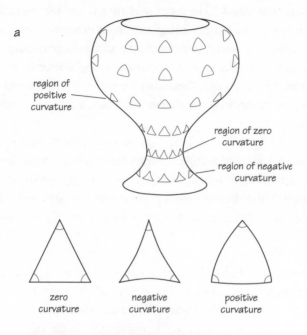

Figure 2.7 (a) Triangles on curved surfaces formed by the shortest distances joining three points. On a positively curved region at the top of the inverted vase, the interior angles of the triangles sum to more than 180 degrees; on a negatively curved region, near the lip, they sum to less than 180 degrees; in between there is a place where the curvature is zero and the angles of triangles sum to 180 degrees. (b) Leaves of kale, a variety of cabbage, have negative curvature.

theologians, scientists or philosophers delving into questions about the ultimate nature of God or the universe were criticised by those who questioned the possibility of ever knowing anything about these things, they pointed to Euclid's geometry as an example of how human thought had grasped a piece of ultimate truth. This is why they sometimes laid out their treatises in the style of Euclid. He was the gold standard.

The discovery that there could be logically consistent geometries on curved surfaces, like saddles or spheres, would have come as no surprise to navigators and artists who had used them intuitively for hundreds of years, but it revolutionised human thinking in unexpected ways. Suddenly there were many possible geometries. Each was just a logical system of rules and starting assumptions. None had any special claim on being part of ultimate truth. Consequently geometry, and all of mathematics, changed its attitude towards systems of axioms and rules. They all 'existed' in the sense that they were self-consistent logical possibilities but this did not endow them with actual or inevitable physical existence.

The simplest examples of non-Euclidian geometries are those which describe surfaces that have negative or positive curvature. As you can see in the picture of the vase here (Figure 2.7a), it is possible for a surface to be quite complicated and have places where the curvature is positive, negative and zero (that is, 'flat'). A simple way to decide what sort of surface curvature you are dealing with is to pick three nearby points, A, B and C, and draw the shortest lines you can from A to B, B to C and then C back to A. On a flat surface those shortest lines will be straight lines and ABC will be a triangle with its three interior corner angles adding up to 180 degrees.

On a positively curved surface, like the surface of a sphere, the shortest possible distances between A, B and C will not be 'straight' in the way they were on the flat surface. They will be circular arcs with centres at the centre of the sphere. These are the 'great circles' that inter-continental flights follow in order to minimise fuel consumption (assuming the winds are low). They close to form a bulging triangle whose interior angles add up to more than 180 degrees. This is the hallmark of positive curvature. Similarly, on a negatively curved surface, like a saddle, a Pringle potato crisp, holly leaves, or leaves of kale (Figure 2.7b),[38] the interior angles of the curved triangle add to less than 180 degrees.

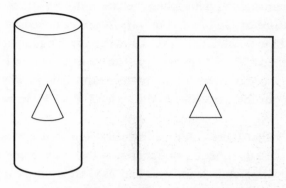

Figure 2.8 A cylinder can be made by joining two opposite sides of a plane square. The cylinder still has a locally flat geometry like the square. A triangle on its surface has interior angles that sum to 180 degrees, just like for the square.

Sometimes being curved doesn't fit exactly with our naïve intuitions. From what we have just seen you might have thought that a cylinder was curved. Not so; if you take a flat rectangle of paper and draw a triangle on it then the three interior angles will sum to 180 degrees, as expected. Now stick the two long sides of the paper together to make a cylindrical tube with the triangle showing on the outside. It will look just the same with its three interior angles summing to 180 degrees. The surface of the cylinder is not locally curved in our sense (Figure 2.8).

Karl Schwarzschild was a genius who never lived to see the real importance of his ideas. He died, aged only forty-two, in March 1916. He made many discoveries in the study of stars, galaxies and gravitation, found the precise description of the black holes that populate our universe today, and paved the way for all the precision experimental tests of Einstein's revolutionary theory of relativity. Yet prior to all that, in 1900, he presented a new picture of the astronomical universe that sprang from the recent understanding of curved geometries. In a lecture at the meeting of the German Astronomical Society in Heidelberg in July 1900, Schwarzschild proposed that the geometry of the universe was not flat like Euclid had taught us, but might be curved like the non-Euclidean geometries first envisaged by Johannes Lambert and the Italian Jesuit mathematician Giovanni Saccheri, in the early eighteenth century, and then developed in more detail by

Figure 2.9 Karl Schwarzschild (1873–1916).

Riemann, Gauss, Bolyai and Lobachevskii[39] in the early nineteenth century.[40] These new possibilities were not universally welcomed by physicists and astronomers and even a far-sighted physicist like James Clerk Maxwell referred to the proponents of these geometrical exten-sions as 'space crumplers' in a postcard[41] to his long-time[42] Scottish friend Peter Tait, in 1874.

Schwarzschild first realised that if the universe possessed a negative curvature then there would be a minimum parallax angle for stars, as had been pointed out first by Lobachevskii, so he deduced that the radius of curvature of space must exceed sixty light years. More interestingly, he then moved on to consider the situation if the universe was positively curved. This meant that it would be finite but unbounded, like the surface of a sphere, closed in upon itself.[43]

He found that the 100 stars with measured parallaxes, together with the 100 million stars whose parallaxes were too small to measure (extending less than 0.1 seconds of arc), could be accommodated satisfactorily without overcrowding in a spherical space of positive curvature whose radius was no smaller than 2500 light years. He also noted that in such a space, if we look in the opposite direction to that of the Sun, we could in principle[44] 'see' it because light rays go all the way around the sphere before reaching our eye.

At the outbreak of the First World War, Karl Schwarzschild volunteered for military duty and during his service in Russia he wrote two remarkable research papers on quantum theory[45] and on Einstein's relativity theory, both of Nobel Prize-winning quality. Unfortunately, in 1916, he developed pemphigus, a severe skin disease caused by an immune system breakdown, for which there was no known treatment. He was taken home in the March of that year but died just two months afterwards.

Here endeth the old world view. The nineteenth-century conception of the universe had shied away from novelty until the very end and Schwarzschild's ideas attracted little attention at the time. Only two cards continued to be shuffled, offering a choice between a universe filled with other galaxies and one in which the Milky Way is the only one and all the distant nebulae lie within it. The ancients would have recognised the alternatives. But the reach of the human mind was about to expand in a very big way.

3 Einstein's Universes

COMPLETING A COPERNICAN VISION

All the pictures of the universe drawn by nineteenth-century astronomers used the style guide that Newton first provided in 1687. His
famous laws of motion and gravity are useful for all sorts of practical
purposes – building bridges, testing cars, guiding aircraft, throwing
stones – but when you look more closely there is a deep problem
lurking within them. Unfortunately, they only hold for a very special
type of observer – *one who does not rotate or accelerate relative to the most
distant stars.* As we saw in the last chapter on p. 25, if you look out
of the window of a spinning rocket (Figure 3.1) you will see stars
accelerating even though they are acted upon by no forces.
   Einstein saw this as a serious problem in the way our laws of Nature
were being formulated. It was scandalous to him that we could have
a description of natural laws that picked on some special collection
of observers for whom, by virtue of their motion, the world was
going to look simpler: it is tantamount to saying that there are some
people with special revealed knowledge that is not available to everyone

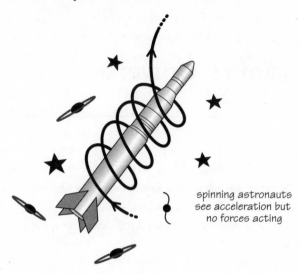

spinning astronauts
see acceleration but
no forces acting

Figure 3.1 Looking through the windows of a spinning rocket, astronauts see the stars accelerating past them, although no forces can be seen to act on them.

else. It is in essence a pre-Copernican world view that gives our *motion*, rather than our location, a special status in the universe.

One of Einstein's great achievements was to find a way of finding and formulating laws of Nature which ensures that all observers will see the same laws no matter how they are moving. His new law of gravity, which supersedes Newton's, is known as the *general theory of relativity* and many believe it to be the most remarkable creation of the human mind. It extends the Copernican perspective from merely dictating that our position in the universe must not be privileged to requiring that the laws of Nature themselves must also be found to be the same by all physicists wherever they are and however they are moving. What does this mean in practice?

Suppose that you look up into the sky and observe that when A happens it always produces the result B. So you have detected a law of Nature that we denote by the equation A = B. You then put all your measuring equipment into a spacecraft and launch it into space. Your spacecraft may be spinning or accelerating in a very complicated fashion. When you make careful observations of A it always looks different because of the effects of your motion, and instead of measuring it to be equal to A, like it was on the ground, you measure it to be A★. Likewise, instead of having an effect B, it always has the outcome B★, where B★ is just the value of B with your new motion relative to the ground taken into account. Einstein's formulation ensures that you find the law of Nature to have the form A★ = B★, no matter what the motion of your spacecraft. The As and Bs have different values but the form of the law relating them A = B or A★ = B★ has the same form in the spacecraft or on the ground. Newton's laws were not like this. The law A = B seen in a non-spinning space-craft would turn into a much more complicated one when the same events were observed from a spinning spacecraft:

$$A★ = B★ + (something\ extra)$$

For Einstein, the completion of the Copernican perspective meant that it must apply to the world of Nature's laws, not just to their outcomes: the planets, stars and galaxies.

EINSTEIN'S INSIGHT

The Universe had to be reinvented, and Einstein's principles were like tiny lamps that illuminated his way. He could make out the contours of a theory of gravitation where all the masses, all the particles, all the energy in the universe contributed to its struc-ture: space-time was bent under the weight of matter-energy.
    Jean Eisenstaedt[3]

In the United States in 1931, Einstein attended the premiere of the silent movie *City Lights* in the company of its star, Charlie Chaplin. There was great cheering from the crowds for the two rather different celebrities and Chaplin reportedly remarked to Einstein: 'They cheer me because they all understand me, and they cheer you because no one understands you.'

Einstein's theory of general relativity has become a byword for unin-telligibility and difficulty: the ultimate intellectual challenge. Certainly, the mathematical language that Einstein used to achieve the Copernican expression of his new laws of motion and gravity had at first been chal-lenging even for him. He confessed to being weaker at mathematics than

Figure 3.2 Marcel Grossmann (*left*), with Albert Einstein, Gustav Geissler and Eugen Grossmann, while they were students in Zurich.

he needed to be. His special talent was for physical understanding, not mathematical wizardry. But if he couldn't do some mathematics he usually knew a man who could. His old student friend Marcel Grossmann was a talented mathematician, at home with the most abstract branches of modern mathematics. Grossmann had also recognised Einstein's remarkable talent for looking deeply into the heart of how nature worked and was keen to help him in any way that he could. In 1912, Einstein took up a professorship in Zurich, in preference to offers from other more prestigious universities, so that he could continue to work closely with Grossmann, who was the professor of pure mathematics there.

Grossmann introduced Einstein to the new maths he needed to express his vision of how gravity shaped the Universe. He showed him that his desire for a 'democratic' way to write the laws of Nature so that they looked the same for everyone could be achieved by writing them down in the language of an esoteric branch of mathematics, called tensor calculus, which guaranteed the universality he sought. Grossmann also introduced Einstein to the profound developments that had occurred in our understanding of the geometry of the complicated curved surfaces that Schwarzschild had started to explore a few years earlier. But why did Einstein need to know all that kinky geometry?

Newton's picture of space was a great fixed stage on which the motions of things like planets and comets were played out. Such things could come and go but space was fixed, changeless and unchangeable, regardless of the matter and motions that inhabited its fabric. Einstein's space was far more malleable. Like a great rubber sheet it would be deformed and shaped by the matter and motion that occurred upon it. In places where there was a strong concentration of mass there would be a strong curvature of space. Far away from any masses, space would be increasingly flat and undistorted. When a body moved between two points it would take the quickest path that it could on the curved surface – that is, the straightest line over it. In the vicinity of a large mass the geometry would be significantly pitted and a passing object that took the quickest path would appear to be drawn towards the centre of the pit. After it passed, the direction of its path would be deflected by the shape of the space. In this way, Einstein could attribute gravitational forces simply to the shape of space. Indeed, there was no need for any forces at all, just curved space.

Yet, at first, this just sounds like another way of saying that there *are* forces. But there is more to it than that. In Newton's old picture with fixed space you could imagine spinning a ball on the fixed stage of space. It would not have any effect on you if you stood on the stage a little way away. Things are very different in Einstein's picture. If space behaves like a deformable rubber sheet then spinning the ball will twist the space around it and if you are standing some distance away you will be pulled around in the same direction as the rotation.[4] This is a real difference.

What Einstein had to do in order to turn this vision into a new theory of gravity was to find some special equations – the new laws of gravity, which would tell us what the shape of space and the rate of flow of time would be for any pattern of mass and energy, whether still or in motion. And they would tell us how those patterns were allowed to change so as to ensure that quantities like energy were conserved. The American physicist John Wheeler once encapsulated Einstein's theory in two sentences: 'Matter tells space how to curve. Space tells matter how to move.'

Einstein wrote these equations in the new mathematical language of tensor calculus that Grossmann had introduced to him. This ensured that they automatically looked the same for observers whichever way they moved – spinning, accelerating, jumping up and down, looping the loop. However their laboratories were moving they would deduce the same laws of gravitation.[5]

Einstein's equations had the beautiful property that a purely mathematical theorem governing the way that geometries of curved surfaces could change became the equivalent to the laws of physics which demanded that energy and momentum be conserved in all natural processes. More remarkable still, if you considered the situation where the masses were small and their motions very slow compared to the speed of light, the distortion caused to the geometry of space would be tiny – and Einstein's equations just turned into those describing Newton's old theory of gravity.

A DIGRESSION

Mistakes are good. The more mistakes, the better. People who make mistakes get promoted. They can be trusted. Why? They're not dangerous. They can't be too serious. People who don't make mistakes eventually fall off cliffs, a bad thing because anyone in free fall is considered a liability. They might land on you.
James Church[6]

In 2000, there was a great competition in Great Britain. The public was invited to vote for the greatest Briton of the past thousand years. Celebrity TV presenters and newspaper reporters struggled to persuade voters to rank William Shakespeare ahead of Princess Diana, and

David Beckham behind Charles Darwin. One quality newspaper considered the case of Newton. It was very strong – rainbows, motion, gravity, calculus and all that – but the journalist was unpersuaded. Einstein showed that some of Newton's theories were definitely 'wrong', he wrote.

This commentary betrayed a subtle misunderstanding about how modern science progresses. When Einstein's theory of gravity came along and was confirmed in increasing detail by experiments, it didn't mean that Newton's theory was put on the scrapheap. Eventually, Einstein's theory will be superseded as well, yet we won't discard it either.

Einstein's ideas extended Newton's theory so that we could understand what happened when gravity became very strong and motions approached the speed of light. Newton's theory could not cope with such extreme situations. But when you look at the form of Einstein's theory in the limiting case that the motions become much slower than light-speed and gravity is weak, then it looks increasingly like Newton's theory, in that Newton's theory is a limiting approximation of Einstein's theory. It supersedes Newton's theory in the scope of its application and the range of situations to which it can be applied, but it does not replace it as a valid description of the everyday world where motions are slow and gravity is weak. This is why students of engineering study Newton's laws and I confidently predict they will still do so in much the same way for a thousand years to come.

A similar view can be taken of the way in which the quantum mechanics of Werner Heisenberg and Paul Dirac superseded Newton's mechanics. Quantum mechanics extended the domain over which we understand mechanics down to very small distances and intervals of time, and to the motion of very low-mass particles and photons of light. Yet, as we let things become big in an appropriate way, [7], quantum mechanics looks increasingly like Newton's mechanics.

In the evolution of our theories of physics, the new theories subsume the successes and insights of their predecessors. They enlarge their domain of applicability and the range of situations they can accurately describe. New theories do not overthrow old theories and consign them to the dustbin of history. Maybe that was once possible, long ago, when the existing theory had very little evidence to support it and negligible explanatory power. Today, a new theory needs to explain all that has already been well explained and something of what hasn't, and have an ability to predict something that no one has thought of before.

THE CREATION OF EINSTEIN'S UNIVERSE OF MOTIONLESS MATTER

People are wrong when they say opera is not what it used to be. It is what it used to be. That is what's wrong with it.
  Noël Coward[8]

Einstein announced his new theory of gravity to the world of science on 25 November 1915 in an article published in the august *Proceedings of the Royal Academy of Sciences of Prussia*. It had taken him more than ten years to solve to his satisfaction the problem of gravity and deduce a set of equations that could always find how mass and energy would curve space and move under its orchestration. His theory predicted exactly the mysterious 43 seconds of arc per century wobble in the orbit of the planet Mercury that had challenged astronomers for an explanation ever since the French astronomer Urbain Le Verrier discovered it in 1859. Newton's theory could not easily explain it.[9] For Einstein it was the synthesis of mathematical formulation and physics – the way the laws of physics emerged almost uninvited from abstract mathematics – that excited him enough to write to one of his friends that 'Scarcely anyone who has understood this theory can escape from its magic.'[10]

Eighteen months later, on 8 February 1917, in the midst of the First World War, Einstein announced the first application of his new theory to the universe as a whole. Every solution of his equations described a possible universe. Yet there seemed to be only one universe, so how do you filter out the unwanted possibilities? Einstein struggled long and hard with this question. If he allowed the universe to be infinite he couldn't see how his equations could constrain its behaviour infinitely far away. If it is finite then he had to avoid there being an 'edge' to space.

Following the developments that had caught Schwarzschild's eye, Einstein saw the importance of a positively curved space. It was finite, but like the surface of a ball, so it had no edge. He also believed in symmetry: the universe should be the same in every direction and in every place, on the average. So although the curvature of space would possess little variations all over the place, just like the surface of a calm sea, viewed in the large it would be about the same everywhere and in all directions at any moment. One of the interesting consequences of the curvature of space is that, although things look the same in every direction, this doesn't mean you are at the centre of the universe. If you walk around like an ant on the surface of a sphere then wherever you are things will look the same in every direction but there is no centre on that surface.[11]

But Einstein then faltered from taking a monumental leap. When he used all these simplifying assumptions he couldn't find possible universes that stayed still, all the possible worlds had to change with time, everywhere, expanding or contracting in size. This was something totally unexpected. Space may be curved but for Einstein in 1917 it had to be a static and fixed arena upon which all the stars moved. The only way that he could find an unchanging universe was to introduce into his equations a possibility that he had hitherto downplayed.

Newton's theory of gravity tells us that the gravitational force of attraction between two masses causes them to accelerate towards one another. In order to stop this acceleration there needs to be a counter-effect of repulsion so that the overall acceleration the masses feel is:

Acceleration = − (gravitational force of attraction) + (force of repulsion)

Einstein's theory *allowed* that force of repulsion to exist but didn't require it to: it was an optional extra. It appears to be an add-on that Nature hasn't made use of because there is no evidence for it when we study gravity here on Earth or its effects in the solar system. Its effects get weaker and weaker at small separations between masses. But it gets stronger when their separations increase.[12] This means that there is some distance in the universe at which the repulsive force will become equal to the attractive effect of gravity. For a universe of this special size there is neither expansion nor contraction. This is Einstein's static universe.

Remember that this universe has a positively curved space and so

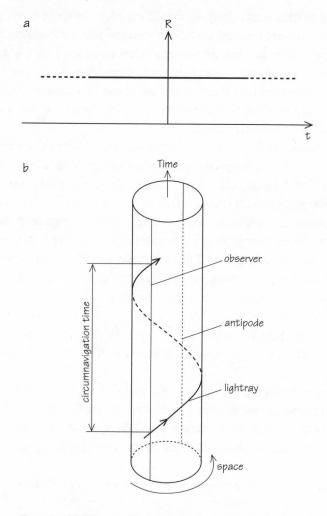

Figure 3.3 (a) Distance, R, versus time, t, in Einstein's static universe. There is no change in the separation of free particles in the universe. (b) A diagram showing the path through space and time of a light ray moving through the space of Einstein's static universe as time passes. It traces out a helix on the surface of a cylinder.

it can be finite in volume yet have no boundary. If we draw a graph of time going upwards and space (two-dimensional slices of it at each moment of time) perpendicular to time's arrow, then the path through space and time taken by someone in Einstein's universe winds around a spiral on the surface of a cylinder (Figure 3.3). If we watch a space-ship moving away from us it first diminishes in size but then comes back and begins to magnify again. The time it takes for light to go all the way around this universe[13] is determined by the average density of matter within it.[14] For a universe of density equal to that of the air around us now, the round-trip time is about 2.5 days. In such a universe we would be able to see what we were doing 2.5, 5, 7.5, 10, etc. days ago, each time the light completes circuits of the cosmos en route to us from the past.

Einstein's universe shows the strength of the inherited conception of a static space that Schwarzschild had also employed. He introduced a striking model universe – a finite curved space with no boundary that exists for all past and future eternity. It was the first to be extracted from his extraordinary equations – but he had suppressed what the equations were trying to tell him: the universe didn't want to be static. Later, Einstein would call his response 'the biggest blunder of my life'.

THE SECOND UNIVERSE: DE SITTER'S UNIVERSE OF MATTERLESS MOTION

I am very interested in the universe – I am specialising in the universe and all that surrounds it.
  Peter Cook

The next scientist to search Einstein's universe equations was a distinguished Dutch astronomer named Willem de Sitter (1872–1934), who benefited from Dutch wartime neutrality to meet and correspond with Einstein.[15] He also kept close scientific contacts with British astronomers, notably Arthur Eddington, who planned the 1917 programme of speakers at the Royal Astronomical Society's monthly meetings in Piccadilly. It was in the third of a series of presentations to the Society that de Sitter revealed a new solution of Einstein's equations.[16]

De Sitter kept Einstein's new repulsive force but decided that he would assume the density of matter in the universe was zero. Of course, the real universe is not empty, but de Sitter was assuming that the density was so low that the effects of the attractive gravity exerted by matter were completely negligible compared to that of Einstein's repulsive 'lambda' force, as it had become known because of the Greek

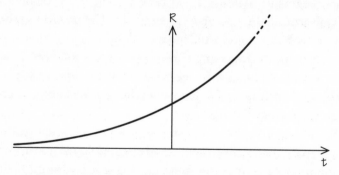

Figure 3.4 De Sitter's accelerating universe. Separations between free particles grow exponentially with the passage of time.

letter lambda (Λ) used to denote it. Unlike Einstein's universe, the
geometry of de Sitter's universe was Euclidean and it was infinite in
extent.

Although de Sitter's universe was very easy to find it was not so
easy to interpret. It seemed to suggest that light from distant objects
would have its wavelength stretched, and so appear redder in colour
and the stretching would become greater as the distance of the source
from us increased. This became known as the 'de Sitter effect'. In 1912,
the American astronomer Vesto Slipher had discovered a significant
shift in an identifiable wavelength of light from a distant nebula (or
'galaxy' as we would now call it); five years later, he reported his
observations of red colour shifts in the light from more than twenty
other nebulae. He had no way of explaining or interpreting these
shifts. De Sitter now showed that his solution of Einstein's equations
created this very effect. Closer scrutiny revealed why: de Sitter's
universe was *expanding*. If you introduced two marker points into it
then they would accelerate apart with their separation increasing
exponentially with time. The change of separation between two reference points with time in de Sitter's universe is shown in Figure 3.4:
their distance apart *accelerates* as time passes.

In this picture of an expanding universe the 'de Sitter effect' has a
simple interpretation. When light waves are emitted from a receding
star those waves have their wavelengths 'stretched' and they are
received less frequently by us than they were emitted. This applies to
all waves, notably sound waves and light waves. When the source

approaches us the opposite happens. A receding light source looks redder, but an approaching one looks bluer. A receding sound wave has lower pitch; an approaching one has a higher pitch. This phenomenon is called the 'Doppler effect' after the Austrian physicist, Christian Doppler, who discovered it in 1842 when seeking an explanation for the different colours of moving stars.[17] In the case of sound it is familiar to us. Recall the noise of the boy racer on his motor bike roaring past your bedroom at 3 a.m. The noise is distinctive: *eee-yowwn*. At first he is coming towards you, the pitch of the engine noise rises (*eee*), then he passes and the sound waves are coming back from a receding source less frequently so the pitch falls (*yowwn*).

Slipher's observations were consistent with light waves reaching us from stars that were moving radially away from us on one side of our galaxy and approaching from the other side. Of course, Slipher could just have been seeing some objects that were drifting past, coming towards us from one direction and receding after they passed. Gradually, observations on both sides of the galaxy showed objects moving away, but Slipher clung to his drift hypothesis to explain the redshifts. There was no reason for him (or anyone else at that time) to think that the whole universe is expanding or even to think about what that might mean.

De Sitter's mathematical universe did describe a universe in which the whole of space was expanding, accelerated by Einstein's new lambda force. Unfortunately, nobody wanted to make the connection with Slipher's observations in 1917, not even de Sitter. By 1921, de Sitter knew that twenty-two of the spiral galaxies that Slipher was seeing were moving away from us. No one knew how far away they were and it was still possible that random local motions were responsible rather than some grand systematic expansion of the universe. De Sitter was reluctant to draw any strong conclusions from Slipher's observations. He had found the first expanding universe but it came heavily disguised. It would be some while before its character was fully revealed.

De Sitter's universe will turn out to be spectacularly important for our current understanding of the universe. Notice some of its features. It keeps getting bigger and bigger for ever and it has no beginning and no end. Although it gets smaller and smaller as you trace it backwards in time, it never reaches zero size and there is no apparent beginning where the size is zero and the density of matter is infinite. The rate at which it expands has a constant value and is always the same. If you are dropped into this universe at some moment of history you have no way to locate yourself in time: the future is indistinguishable from the past. All the things that you can observe are always the same. History is not an important subject in de Sitter's world.

FRIEDMANN'S UNIVERSES OF MATTER IN MOTION

We have Einstein's space, de Sitter's space, expanding universes, contracting universes, vibrating universes. In fact, the pure mathematician may create universes just by writing down an equation and indeed if he is an individualist he can have a universe of his own.

    J. J. Thomson

Einstein was a physicist, de Sitter was an astronomer, but the next famous entrant in the game of finding universes was an unknown young mathematician and meteorologist from Petersburg named Alexander Friedmann.[18] As a young student of physics he was lucky to attend lectures on quantum theory and relativity given by the outstanding Austrian physicist Paul Ehrenfest, who taught at the University in Petersburg from 1907 to 1912 before moving to Leiden. Friedmann kept in contact with Ehrenfest after he graduated and began work as a meteorologist at the observatory in Pavlovsk and then as a research student of the Norwegian Vilhelm Bjerknes (the founder of modern theoretical meteorology) in Leipzig. After his service in ballistics on the Austrian front in the First World War[19] he returned to academic work and made rapid progress, working in mathematics, mineralogy and atmospheric science, eventually becoming a professor of mathematics and physics at Perm State University, a new outpost of Petersburg University, in 1918. There, he suffered the consequences of civil war, with the city of Perm occupied first by the anti-communist 'Whites' and then by Trotsky's Red Army, which led to the departure of most of his colleagues. In 1920,

Figure 3.5 Alexander Friedmann (1888–1925).

Friedmann moved to the Geophysical Observatory in Petersburg, where he began to learn about Einstein's new theory of general relativity. Friedmann's remarkable breadth of expertise extended from purely theoretical work in mathematics to dramatic high-altitude balloon flights to investigate effects of altitude on the human body. Together with a colleague he held the world ballooning altitude record for a period in 1925, ascending to a height of 7400 metres. He died just a few months later, apparently of typhus fever, at the age of thirty-seven.[20]

Friedmann had learnt the formidable mathematics behind Einstein's equations in great detail and set about finding more general solutions than those found by Einstein and de Sitter, while keeping to their assumptions that the universe was the same everywhere and in every direction. His two papers, of 1922 and 1924, and his book *The World in Space and Time*, written in 1923, reveal that Friedmann knew of universes found by de Sitter and Einstein but seems to have been unaware of Slipher's discovery of red-shifted light from distant stars.

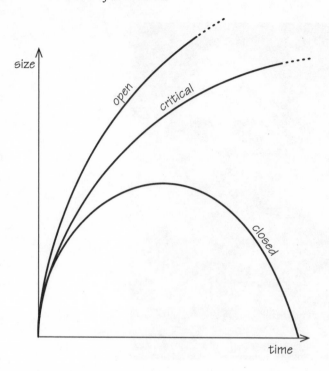

Figure 3.6 Friedmann's expanding and contracting universes.

His approach to Einstein's equations was that of a mathematician looking for solutions. And he found them.

First, he discovered the possibility of a 'closed' finite universe with a space of positive curvature that expands from a beginning at a finite time in the past to a maximum size and then contracts back to an end at a finite time in the future (see Figure 3.6). This was an expanding universe that contained ordinary matter that exerts no pressure. It was finite in mass and volume and finite in total lifetime – Friedmann even estimated that its mass would be about 5×10^{21} times the mass of our Sun if the time-span of an entire cycle lasted about 10 billion years.[21] It began with what would later be called a big bang of infinite density and contracted to its end at a similarly extreme big crunch. In his book, Friedmann reflected on the fact that he could imagine continuing this solution forwards (and backwards) in time so that the universe could oscillate through a never-ending series of expanding and contracting cycles (shown in Figure 3.7), remarking that:

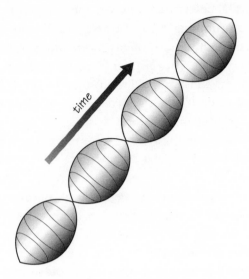

Figure 3.7 Friedmann's oscillating universe.

there can be cases when the world's radius of curvature . . . changes periodically. The universe contracts into a point (into nothing) and then again increases its radius from a point up to a certain value, then again, diminishing its radius of curvature, transform itself into a point etc. One cannot help thinking of the tales from the Indian mythology with their periods of life. There is also the possibility to talk about the creation of the world out of nothing. But all this has at the moment to be considered as a curiosity which cannot be verified by the inadequate astronomical observations.[22]

Next, Friedmann realised that there were also solutions of Einstein's equations in which the curvature of space was 'open' and negative, like a saddle, so that the volume of the universe is infinite, starts expanding from a beginning in the finite past and continues expanding for ever.[23] These are also shown in Figures 3.6 and 3.7.

Thus Friedmann was the first to discover that Einstein's equations allow the possibility of expanding and contracting universes containing ordinary material, like planets and stars. He wasn't spurred on by astronomical observations and he didn't dwell on the physical meaning of the beginning or the end of these universes in time. It is also interesting that he describes a universe appearing out of (and disappearing back into) nothing. Unfortunately for Friedmann, no one seems to have taken any notice of his momentous discoveries. He published his papers in the *Zeitschrift für Physik*, a leading research journal of the day, where Einstein published some of his work, yet there was no discussion of his new 'universes' at all. Worse still, Einstein thought that Friedmann had done his sums wrong and these new expanding and contracting universes were not really solutions to his equations at all, and published a short note to that effect in the journal.

Fortunately, one of Friedmann's St Petersburg colleagues, Yuri Krutkov, travelled to Leiden in May 1923, where he met Einstein and was able to convince him that Friedmann's calculations were correct: there really were non-static universes as solutions to his equations. Einstein acted quickly and published another short note in the journal saying that after correspondence with Friedmann and discussions with Krutkov he recognised that he had made a calculational error and that Friedmann's solutions were 'correct and clarifying. They show that in addition to the static solutions to the field equations there are time varying solutions.'[24] Interestingly, in the handwritten version of this note to the journal he had added a final sentence saying that, despite their correctness, 'a physical significance could hardly be ascribed to these solutions'. Fortunately, he deleted that sentence before his letter was published.

Friedmann didn't have much longer to live and he didn't follow up his discoveries by interpreting them astronomically. For him they were just pieces of mathematics. One of his colleagues, Vladimir Fock, reported that Friedmann had once told him that: 'his task was to indicate the possible solutions of Einstein's equations, and that the physicists could do what they wished with these solutions'.[25] But today no name is more closely linked with the word 'universe' than Friedmann's; if you Google Friedmann's universe, you will get more than a million hits.

LEMAÎTRE'S UNIVERSES

The evolution of the world can be compared to a display of fireworks that has just ended: some few red wisps, ashes and smoke.
Standing on a well-chilled cinder, we see the slow fading of the
suns, and we try to recall the vanished brilliance.
        Georges Lemaître[26]

Georges Lemaître was ordained a Catholic priest by the Maison Saint-
Rombaut seminary in 1922 after receiving the Military Cross in the
First World War. The war had interrupted his studies and he graduated in mathematics after enrolling to read engineering at the Jesuit
University of Louvain in Belgium in 1920. He gained a scholarship to
study overseas and spent the year 1923–4 at St Edmund's House[27] in
Cambridge, working with Arthur Eddington as a visiting student at
the University Observatories. Eddington was arguably the most accomplished astrophysicist in the world at the time, with a remarkable list
of achievements to his name – understanding the way stars work, the
development of theories of star motions in our Galaxy, and the lead-
ership of a famous expedition in 1919 to the Portuguese island of
Principe, off the coast of West Africa, in order to test Einstein's prediction that light rays from distant stars would deviate from straight lines
because of the effect of the Sun's gravity. He had also written the first
advanced text in English explaining Einstein's general theory of
relativity.

Eddington's mathematical ability was legendary and he had very
quickly developed a deep understanding of Einstein's theory of gravity.
He had also played an important role in maintaining scientific links
with continental scientists during the First World War through his
position as Secretary of the Royal Astronomical Society and, as a
Quaker of deep conviction, his non-participation in the war.
Remarkably, Eddington was saved from the social disgrace and probable imprisonment that would have resulted from him claiming
conscientious objection to military service in 1917 by the intervention
of the Astronomer Royal, Sir Frank Dyson. Using his close connection
to the Admiralty, Dyson brokered an agreement whereby Eddington's
military service would be deferred; instead, if the war had ended by

then, he would lead one of the two expeditions planned by the Admiralty to observe the total eclipse of the Sun on 29 May 1919 in order to test Einstein's general theory of relativity.

Lemaître didn't study cosmology with Eddington when he spent that year as a visiting student in Cambridge but he took the opportunity to develop a profound understanding of general relativity. After Cambridge he went to work on his Ph.D. with the famous American astronomer Harlow Shapley at Harvard Observatory and graduated from the neighbouring Massachusetts Institute of Technology in July 1927 (the Observatory did not start awarding Ph.D.s until 1929). Eddington had been enormously impressed with Lemaître's brilliance and mathematical ability and recommended him strongly in letters to other scientists. Lemaître was also an extremely gregarious and friendly individual who got on well with everyone he encountered in his scientific travels. This undoubtedly encouraged collaboration and exchange of ideas.

During his stay in Boston, Lemaître acquired a thorough understanding of the problem of the redshifts and he had already read the early papers of Einstein on the static universe model. Neither he nor Eddington had seen Friedmann's work. By 1927, Lemaître had prepared the most complete study of the simplest universes predicted by Einstein's theory. It went further than Einstein, de Sitter and Friedmann by introducing the possibility that the universe contained radiation with significant pressure as well as stars and galaxies. He also sought to account for the redshifts first seen by Slipher by means of the Doppler shift in an expanding universe.

Lemaître's impressive paper of 1927, published first in French in an obscure Belgian journal, is the first that combines the expanding universe solutions of Einstein's equations with their physical interpretation and a calculation of the redshift of light from distant stars as a Doppler effect.[28] As in all his work, Lemaître displays a wonderful clarity, using no unnecessary mathematics, yet capturing all the essential points of physical significance.[29] He appreciated that the universe has no centre and no edge, that it can be finite or infinite, and that Einstein's equations have a simple interpretation in terms of energy conservation and thermodynamics. He even calculated the present expansion rate of the universe from the observational data on the redshifts and distances of forty-two galaxies and gave the first determination of the so-called 'Hubble constant', H (getting a value of 625 km per sec per Megaparsec,

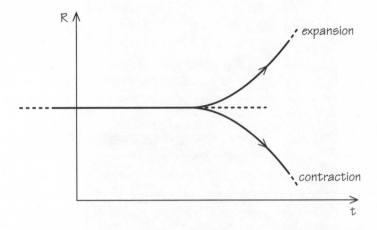

Figure 3.8 The instability of the static universe. Lemaître and Eddington both showed that the slightest movement of matter in a static universe produces a state of future expansion or contraction.

similar to that found by Hubble two years later). He calculated the ratio of their recession speeds (v) against their distances (r)[30] and gave the first deduction of 'Hubble's law' ($v = Hr$) using the Doppler effect.

Hubble would publish his own evidence for this law in 1929 and de Sitter followed up with his analysis of the same data (to Hubble's great annoyance because he regarded the data as his own even though it was published and de Sitter referenced its source) in 1930. Hubble never really subscribed to the overall physical implications of an expanding universe or used his observations specifically to support theoretical models. He would refer to the velocities of distant galaxies as 'apparent' and left the interpretation to others.[31] De Sitter and Lemaître, however, took up the expanding universe picture with great enthusiasm.

The status of Einstein's static universe and de Sitter's exponentially expanding empty world were also made clearer by Lemaître's work. What Lemaître had shown was that Einstein's world was *unstable*. If the universe began in a static state then any disturbance or motion within it would launch it into a gradual state of expansion or contraction (Figure 3.8). It was the cosmological equivalent of a needle balanced on its point.

This discovery, which Einstein does not seem to have greeted with enthusiasm, was the catalyst for Lemaître's universes to become well

Figure 3.9 Georges Lemaître and Albert Einstein in 1933.

known. In 1930, his old mentor, Arthur Eddington, had suspected that Einstein's static universe was unstable to small changes and proved it by introducing small density irregularities into Einstein's solution, and then showing that they would all grow. He was oblivious to Lemaître's work, but soon after he published[32] his discovery he was shocked to receive a letter from his former student pointing out that the instability of Einstein's universe had been demonstrated (in a different way) in his 1927 paper. Eddington had forgotten about that part of the paper and not remembered its significance for the calculations he had done. He acted quickly, writing a letter to the journal *Nature* in June 1930 drawing attention to Lemaître's neglected work and arranging for an English translation of the 1927 article to appear in the *Monthly Notices of the Royal Astronomical Society* in 1931.[33] As a result, Lemaître became the best-known theoretical cosmologist of the day. He only found out about Friedmann's earlier mathematical work at a conference in October of that year, but he added a citation to it in the 1931 translation of his paper.

Figure 3.10 Lemaître lecturing on celestial mechanics.

In 1957, two years after Einstein's death, Lemaître gave an interview about his meetings with Einstein and revealed that at a 1927 conference in Solvay, during private conversations, Einstein had commended the mathematical elegance of his paper, telling him of Friedmann's earlier work, but from a physical point of view he regarded such non-static cosmologies as 'abominable'.[34] Lemaître still felt that Einstein had not really absorbed the significance of the new astronomical observations of receding galaxies for the expanding universe solutions of his theory. However, by 1933, after hearing Lemaître lecture on the subject in Pasadena, Einstein became convinced of the simplicity of Lemaître's approach and described his picture of a universe expanding from a hot beginning as the most 'beautiful' explanation of the universe's behaviour.

Lemaître's universe was similar to the one which Eddington discovered in his study of the instability of Einstein's static universe. It is often called the Eddington–Lemaître universe (Figure 3.11) and begins in the infinite past in a static state which then gradually expands so

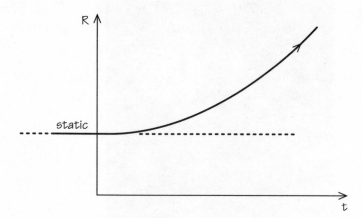

Figure 3.11 The Eddington–Lemaître universe.

that the expansion becomes noticeable at some finite time in our past.[35] It then continues to expand and becomes increasingly like de Sitter's exponentially expanding universe after a long time. It has Einstein's repulsive lambda force; its space is positively curved and finite in extent, and it always expands.

The infinite past age means the universe has no beginning. Neither Eddington nor Lemaître seemed concerned by this. Indeed, Eddington regarded it as a natural feature, giving the universe 'an infinite time to get started', because 'There is no hurry for anything to begin to happen'.[36] He found a universe with a sudden beginning in time, like Friedmann's, 'repugnant', a mere 'fireworks theory' in contrast to his own 'placid' conception that was more likely to satisfy our general sentiments for tranquillity. Eddington also argued that although there was an infinite past history there would only be a finite past during which things happened because the universe was too close to complete thermal equilibrium for there to be any significant accumulation of entropy from its infinite past (Figure 3.13). Thus, he believed that this type of universe would not encounter a heat death brought about by an unacceptably large accumulation of entropy and disorder over its infinite past history. It was geometrically old but thermodynamically young.

One might have expected that Lemaître's religious convictions would have prejudiced him towards models of the universe which had a beginning in time. Yet this was not the case. He kept his

Figure 3.12 Arthur Eddington and Einstein in 1930, talking in the grounds of the Cambridge University Observatories, where Eddington lived with his sister.

scientific and religious views separate, seeing them as having no points of possible connection or conflict; they were parallel but different interpretations of the world. For him, the Bible taught no science and to seek religious instruction from science was like looking for Catholic dogma in the binomial theorem.[37] Later in his life, when President of the Pontifical Academy of Sciences, Lemaître wrote of his theory of the expanding universe that:

As far as I can see, such a theory remains entirely outside any metaphysical or religious question. It leaves the materialist free to deny any transcendental Being . . . For the believer, it removes any attempt at familiarity with God . . . It is consonant with Isaiah speaking of the hidden God, hidden even in the beginning of the universe.

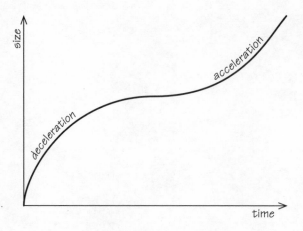

Figure 3.13 Lemaître's universe begins expanding from a big-bang beginning like one of Friedmann's universes, coasts near the static universe before starting to accelerate, and becomes increasingly like de Sitter's universe.

None the less, he seemed to favour one of the many universes that he found in his 1927 paper. It did have a finite past and expanded from a hot dense beginning, first decelerating before gradually switching to accelerating expansion as the repulsive cosmological constant came to dominate over Newton's gravitational force of attraction, before continuing on towards an exponentially fast expansion like de Sitter's universe (Figure 3.13). It had a positive space curvature and positive cosmological constant, just like Einstein's static universe, but the positive repulsion is slightly larger than the very special value chosen by Einstein and so this universe always expands.

Lemaître's universe has turned out to be the most accurate description of our universe, with a total age of 13.7 billion years and the transition to acceleration occurring about 4.5 billion years ago.

The clear analysis provided by Lemaître made it straightforward for cosmologists to survey the entire gallery of simple universes which expand at the same rate everywhere and in every direction. There were only two quantities that could be changed: the curvature of space could be positive, negative or zero (Euclidean) and the cosmological constant force introduced by Einstein could be repulsive (positive), attractive (negative) or zero. In Figure 3.14 is a gallery of

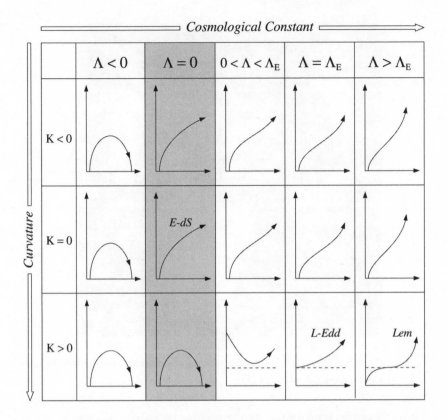

Figure 3.14 The gallery of possible Friedmann–Lemaître universes, showing the possible changes of distance with time for all the possible combinations of values for the space curvature (K positive, zero or negative) and the cosmological constant. The special value of the cosmological constant, denoted by Λ_E, is the value that gives the special case of the Einstein static universe (shown dotted) when the curvature is positive. This static universe is unstable and eventually starts to expand or contract, as shown in the frames *L-Edd* and *Lem*.

all the possible universes that can be obtained, which were first tabulated in this form by Edward Harrison in 1967.[38]

Figure 3.15 Einstein and de Sitter working together on their universe model at Cal Tech, Pasadena, in 1932.

THE UNIVERSE OF EINSTEIN AND DE SITTER

I do not think the paper very important myself, but de Sitter was keen on it.
   Albert Einstein

You will have seen the paper by Einstein and myself. I do not myself consider the result of much importance, but Einstein seemed to think it was.
   Willem de Sitter[39]

In the early spring of 1932, Einstein and de Sitter joined forces to publish a short two-page note which attempted to simplify the study of cosmology. [40] The gallery of possibilities that Lemaître's work had revealed led to many possible expanding universes, some of which kept on expanding for ever whilst others, like those first found by Friedmann, eventually turned around and contracted. Einstein and de

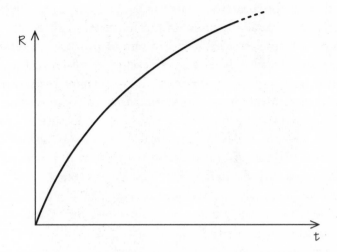

Figure 3.16 The Einstein–de Sitter universe. Distances increase in proportion to the two-thirds power of the time.

Sitter were both visiting the California Institute of Technology in Pasadena at that time (Figure 3.15 shows them at work together there), and pointed out that there was a simplest-possible universe amongst the profusion of spatially uniform and isotropic possibilities.

If the curvature of space is taken to be zero (so the geometry of space is Euclidean), the cosmological constant is set to zero – Einstein was anxious to see the back of his earlier invention – and the pressure of matter is taken to be zero as well, then a very simple universe results. The Einstein–de Sitter universe expands from a beginning in the finite past and expands for ever (Figure 3.16)[41].

This was a very simple deduction from the work that had already been done and it might well not have been judged original enough for publication had it not had such a famous pair of authors. In fact, this model is unstable, just like Einstein's static universe, in the sense that if the curvature is not exactly zero then the expansion will gradually peel away from the Einstein–de Sitter trajectory towards faster runaway expansion or slow down and reverse into contraction, as can be seen in Figure 3.6. The Einstein–de Sitter universe is the middle trajectory shown there and the other closed and open universes with zero lambda force peel off from it as time increases. The novelty of Einstein and de Sitter's suggestion, which neither thought very important, was that for the next sixty years this simple model was the best description of the overall expansion of our universe. The fact that the universe was still so close to this special rate of expansion indicated that the instability hadn't had time to develop significantly. But the universe had been expanding for more than 13 billion years, which suggests that the universe must have started expanding extraordinarily close to the special state of Einstein–de Sitter. This odd state of affairs later became known as the 'flatness problem' and was one of the motivations for the famous inflationary-universe theory proposed by Alan Guth in 1981. We shall encounter it in later chapters.

All these universes have an illuminating and simple interpretation that would have been readily understandable to Newton. If you throw a stone in the air it will return back to Earth: it doesn't have enough energy of motion to escape the gravitational pull of the Earth's gravity. But if you could throw it faster than 11 km/sec then it would not come back.[42] The entire universe has an analogous 'escape speed' which determines whether or not it keeps expanding for ever. The closed, recollapsing universes correspond to the situation where the expansion speed is less than the universal escape speed; the open, negatively curved universes have an expansion speed that exceeds their escape speed. The Einstein–de Sitter universe, with zero curvature, expands at exactly the escape speed and so just manages to expand for ever. Any increase in the density of material it contains, no matter how small, would cause it to expand slower than its 'escape speed' and collapse back in the future.

TOLMAN'S OSCILLATING UNIVERSE

> What goes up must come down.
> Frederick A. Pottle[43]

Friedmann's pioneering work had found the first universes that expand to a maximum size before contracting back down to zero size, as pictured in Figure 3.7. He had alluded to the possibility that there might be a continuous sequence of cycles passing through successive maxima and minima of zero size, like a bouncing ball. Friedmann, as we stressed, was not so interested in the astronomical situation; he just wanted to find mathematical solutions to Einstein's equations. Yet there were

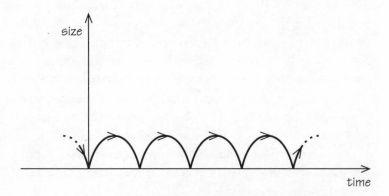

Figure 3.17 An oscillating universe with identical cycles in which each collapse to a big crunch is succeeded by a new expanding cycle, and so on, for ever. There is no beginning and no end to this universe.

interesting things that could be asked about a cosmological scenario in which a closed universe oscillated for ever from a past to a future eternity. Would all the cycles really be the same, as in Figure 3.17?

This question was taken up in 1932 by Richard Tolman at Cal Tech, in Pasadena, an institution that was often visited by Einstein. Tolman had a rather different scientific background to our other pioneering cosmologists. He was a professor of physical chemistry and mathematical physics at Cal Tech and had a special interest in thermodynamics. He considered what would happen if the famous second law of thermodynamics were applied to the solutions of Einstein's equations which described a universe oscillating through successive cycles (Figure 3.18). The increase in entropy from cycle to cycle would be like transferring energy from ordered forms, things like atoms and galaxies, into disordered heat radiation. It is simple to introduce a transfer like this into Einstein's cosmological equations and it means that each successive cycle of an oscillating universe contains a larger fraction of its energy in the form of radiation (which has positive pressure) than its predecessor. This ensures that the maximum size of each cycle is bigger than the last (Figure 3.19). The oscillations grow in size and total age and the universe eventually looks like the Einstein–de Sitter universe for increasingly long periods of time before it ceases expanding and collapses.

If we run this universe backwards in time it may need no begin-

Figure 3.18 Richard Tolman explains the thermodynamics of his oscillating-universe scenario to Einstein after a lecture. The equations on the blackboard behind them describes the thermodynamics of matter and radiation in an expanding universe.

ning. Eventually, if we bounce back far enough into the past, it must have been so small that quantum effects dominate the force of gravity and Einstein's equations may no longer apply. It might also be claimed that Einstein's equations don't hold good all the way down to zero size of the universe at each bounce because the density of matter and radiation would be infinite there. However, if some new quantum gravitational physics intervened to make the bounce occur at a small, but non-zero, radius the same sequence of growing oscillations might still occur.

Tolman was a cautious scientist and many of his investigations of unusual universes were meant as warnings to the unwary who might be tempted to draw premature conclusions about the nature of the universe. Thus, the conclusion he draws from this bouncing universe of growing size and entropy is simply that:

At the very least it would seem wisest, if we no longer dogmatically assert that the principles of thermodynamics necessarily require a universe which

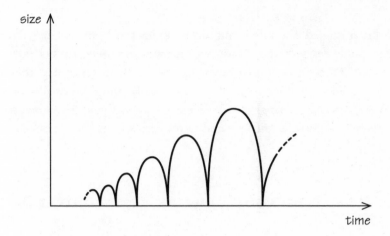

Figure 3.19 Tolman's oscillating universe. The increase in the total entropy of the universe and the conservation of its energy require successive cycles to grow in maximum size.

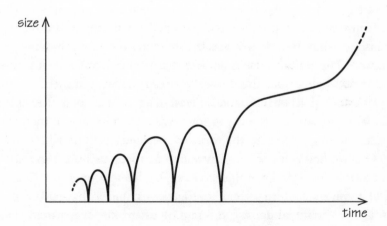

Figure 3.20 The universe of the author and Mariusz Dąbrowski has a positive lambda force. No matter how small the value of lambda, Tolman's oscillations will eventually end and the universe will then expand for ever, becoming increasingly like de Sitter's universe.

was created at a finite time in the past and which is fated for stagnation and death in the future. [44]

Tolman missed one interesting detail about this bouncing universe that was only discovered by Mariusz Dąbrowski and myself much later, in 1995.[45] If Einstein's repulsive cosmological constant is included, then, no matter how small its value, it will eventually bring the oscillations to an end and the expansion will accelerate away towards the behaviour seen in de Sitter's universe and never contract and oscillate again (Figure 3.20).[46]

LEMAÎTRE AND TOLMAN'S KINKY UNIVERSE

– but who could feel at home astraddle
An ever expanding saddle?
    W. H. Auden[47]

In 1933, Lemaître and Tolman both started to think about finding mathematical universes that were more realistic than any that had been found so far. Following Einstein's lead, all the early attempts to find new universes described by his equations had made the simplifying assumption that the universe was the same in every place and in every direction. The real universe is not like that. You couldn't exist if it was. Still, it was hoped that the deviations from perfect symmetry might be small enough to be ignorable, at least in a first attempt at describing the real universe. Yet, when you look out into the universe there are significant irregularities in the form of galaxies and stars all over the sky. Can we find solutions of Einstein's equations that are like that?

Lemaître and Tolman both succeeded, by different means, in finding the first universes which were inhomogeneous; that is, they had properties, like the material density and rate of expansion, that varied from one place to another as well as from one time to another. Lemaître's interest in this type of universe was very far-sighted. He wondered if it might provide a way to understand why there were lumps, like stars and galaxies, in the universe at all. Remarkably, he is able to calculate the behaviour of regions that are denser than average, showing that they become more pronounced over the passage of time (because they can attract even more material at the expense of the sparser than average regions).[47] Hence, a lumpy universe becomes lumpier as it gets older.

Tolman, less speculative and adventurous than Lemaître, worried about strong conclusions being drawn about the universe by using unverifiable assumptions. In particular, all the studies using Einstein's equations had so far assumed the universe was the same everywhere. He thought that was rather unlikely, so conclusions drawn about whether the universe had a beginning, or would collapse to some future end, were rather precarious speculations. They rested upon the pure assumption that the part of the universe we couldn't see was the same as the part that we could see.

When Tolman was able to find solutions to Einstein's equations that described simple irregular universes, he found that they allowed for universes in which different parts behaved like different varieties of the Friedmann–Lemaître universes. The curvature of space and the density of matter could vary smoothly from place to place. In one region the density might be larger than average and the curvature of space positive: this region would behave like a closed universe, expanding to a maximum extent before collapsing back under its own gravity, perhaps to form a galaxy. In another, sparser than average, region the curvature would be negative and the material there would never be able to overcome the expansion and collapse. Tolman suggests that this leads us:

to envisage the possibility that regions of the universe beyond the range of our present telescopes might be contracting rather than expanding and contain matter with a density and stage of evolutionary development quite different from those with which we are familiar. It would also appear wise not to draw too definite conclusions from the behaviour of homogeneous models as to a supposed initial state of the whole universe.[48]

His caution about not drawing conclusions about the beginning was motivated by three features of these kinky universes that Einstein's equations allowed. First, the variation from place to place meant that the beginning of the expansion need not occur simultaneously every-where. Second, it was appreciated later that there could even be an extreme situation where the beginning of some parts of the universe lagged so far behind that of others that the late beginners could see and be influenced by the part that began earlier.[49] Finally, in situations where the cosmological constant was included, there might be some regions where there was no beginning (because they behave locally like the Eddington–Lemaître universe), whereas other regions had a beginning.

Similar differences could arise from place to place regarding the likely final state of the universe and the question of whether the universe is infinite or finite. Tolman pointed out that we could be living in a large, over-dense, positive-curvature region of an infinite universe which will expand for ever, even though our observable part behaves like a finite, closed universe which will eventually contract. If we assumed that the whole universe has the same characteristics as our local patch then we would draw quite the wrong conclusions about its ultimate properties.

MILNE'S UNIVERSE (AND NEWTON'S UNIVERSES)

> No effects of expansion – no recession factor – can be detected.
> The available data still favour the model of a static universe
> rather than that of a rapidly expanding universe.
>   Edwin Hubble [50]

By the mid 1930s, there was a good understanding of the simple types of expanding universe. Edwin Hubble had enlarged the observational store of redshifts and provided stronger evidence that the speed of recession of distant galaxies was proportional to their distances from us. He confirmed Lemaître's first deduction of Hubble's law in 1927, but seemed reluctant to accept that his data confirmed that the universe was expanding. The general behaviour of the simple, homogeneous and isotropic cosmological solutions of Einstein's equations was also well understood. This understanding was helped, and spread to less mathematically minded astronomers, by the work of two English cosmologists, William McCrea (1904–99) and Arthur Milne (1896–1950), who showed that the behaviour of all the universes found so far by solving Einstein's formidable system of equations arises more simply in Newton's old theory of gravity without using Einstein's equations at all.[51] They described the behaviour of a ball of expanding matter subject to Newton's famous inverse-square law of gravity. If you wanted to include the repulsive effects of Einstein's cosmological constant then you just added this extra lambda force to Newton's inverse-square law for the gravitational force between two masses. Universes expanded for ever if their energy of motion was greater than the energy of gravitational attraction between the particles of matter. But if the gravitational attraction won out, then the expansion would reach a maximum and then contract. In between these two situations was the universe of Einstein and de Sitter – exactly poised with the energy of expansion equal to the energy of gravitational attraction, just like a rocket launched from the Earth with exactly the 'escape speed' needed to escape the Earth's gravitational pull. This way of looking at Einstein's simplest universes is still the commonest way of teaching them to students in physics and astronomy.

Milne was a brilliant young astrophysicist who made major contributions to our understanding of the stars and their atmospheres. When he gained an entrance scholarship to Trinity College, Cambridge, he received the highest marks ever achieved by any candidate. He became a Fellow of the Royal Society whilst still in his twenties and was appointed Professor of Mathematics at Oxford in 1928. His interests turned to cosmology in 1932 and he developed a distinctive theory of relativistic motion; he also challenged the widely held notion of the heat death of the universe.

Milne showed that the expanding universe and Hubble's law of recession could be understood without Einstein's theory of gravity. This led him off in many directions to discuss whether the 'time' experienced by atoms need be the same as that experienced by 'gravitational' systems. Yet the new type of expanding universe he favoured could be described very simply using Einstein's equations. The Milne universe, like de Sitter's, contains no matter; but, unlike de Sitter's, it doesn't contain the repulsive force of the cosmological constant either. It possesses only negative spatial curvature and it expands for ever at the speed of light (Figure 3.21).[52]

Milne's universe is the simplest possible universe with the assumption that the universe is uniform in space and isotropic. Although it sounds special and unrealistic (the real universe, after all, isn't empty of all matter and radiation), it has remained of enduring interest because if you add ordinary matter or radiation to an open universe

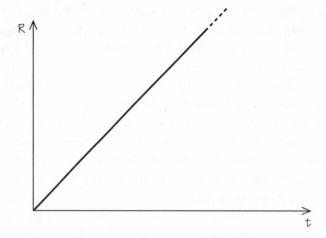

Figure 3.21 Milne's universe. Distances increase in direct proportion to the age of the universe.

with no cosmological constant it will expand for ever and get closer and closer to the behaviour of the Milne universe.[53]

Milne was a scientist with strong (Anglican) religious beliefs and motivations who was often interested in reconciling the scientific picture of the expanding universe with the Christian conception of God and the created universe. He wrote quite extensively on this issue, notably his book *Modern Cosmology and the Christian Idea of God*,[54] published posthumously[55] in 1952. Yet he never wrote for the general public and his books are only easily accessible to physicists and mathematicians. He developed a strong preference for an infinite universe because it offered scope for an unlimited number of evolutionary experiments to be carried out by the Deity. A finite universe he saw as limiting the scope of Divine action.

Milne's mathematical cosmology sought to deduce the one possible 'rational' structure for the geometry of the universe and he believed that this rational structure did place some constraints upon the type of universe God could have created. The universe could not, he believed, expand at different rates in different directions – although, as we shall see in the next chapter, others disagreed with him about that. It must have universal laws – the same everywhere. For he maintained that the laws of Nature were rationally determined by the structure of the universe: the law of gravity could not be very slightly changed without destroying the harmony of the whole cosmos. Thus, for Milne, God was constrained to some extent by logic, which dictated the form that the universe and its laws could take.[56] Yet this was balanced by the freedom introduced by the infinity of space: Milne believed that all biological possibilities could be fully exploited in that infinite cosmos. Curiously, this was the complete opposite to the way Lemaître thought about the same issues. He was strongly attached to the finiteness of the universe. For him, the universe was here for living beings to explore and understand. An infinite universe did not allow that: it was greater than human comprehension, anything could happen somewhere: it was a home too large for its inhabitants.

4 Unexpected Universes: the Rococo Period

Like the ski resort full of girls hunting for husbands and husbands hunting for girls the situation is not as symmetrical as it might seem.
   Alan Mackay

FRACTAL UNIVERSES

The grandeur or infinity of the . . . cosmos added nothing to it. It was like telling a prisoner in Reading gaol that he would be glad to hear that the gaol now covered half the country.
   Gilbert K. Chesterton[1]

Throughout the 1920s, Einstein's theory led to the creation of the modern subject of cosmology – the study of whole universes. Indeed, universes were becoming as common as number 12 buses. Yet, at the same time, some scientists were fighting a rearguard action, trying to convince Einstein that describing universes using Newton's old theory was not the impossibility that Einstein had claimed.

The main difficulty was the problem of dealing with an infinite space that was uniformly filled with matter. Either all the matter in the universe had to be confined within a large island of finite mass or Newton's famous inverse-square law for the fall-off of force of gravity with distance must fail beyond some great distance. In an infinite universe of constant density you could 'prove' that the gravitational force you should feel from all the matter in the universe could be equal to anything![2] Clearly something was wrong.

In the late nineteenth century many cosmological models, like

Schwarzschild's universe with a spherical topology which we met earlier, had grappled with this paradox. In 1907, the Irish scientist Edmund Fournier d'Albe (1868–1933) published an intriguing little book for a wide audience entitled *Two New Worlds*, in which he suggested that it would be natural to think of the universe as having a hierarchical structure, all the way up from atoms to the solar system and beyond, without limit.[3] In effect, his picture of the astronomical universe was one of clusters of clusters of clusters . . . *ad infinitum*, a picture that we have seen suggested before in the work of Wright, Kant and Lambert. It was also impressionistically re-created by the great American writer Edgar Allan Poe, in his 1848 prose poem *Eureka*, which talked of the universe as a never-ending series of 'clusters of clusters', each with different laws and having no contact with us.[4]

One of Fournier d'Albe's motivations was to provide a solution to the problem of formulating a model of the universe with infinite mass. Although his ideas were off-beat he was taken fairly seriously. He was an expert on electricity and magnetism and in 1923 he was responsible for transmitting the first television picture (of King George V) from London; he also invented the optophone, which enabled the blind to 'see' by converting optical signals into sound.

Fournier d'Albe's book was read with particular enthusiasm by a Swedish astronomer, Carl Charlier (Figure 4.1), who used it to develop a more elaborate picture of a Newtonian universe that escaped the paradoxes introduced by infinite space. Charlier introduced a mathematical description of a never-ending hierarchy of clusters (Figure 4.2), organised in such a way that the average density of the entire infinite universe was zero![5] This solved the long-standing problem of explaining why the sky was dark at night: the integrated contribution of all the stars in his infinite hierarchy was negligible.

The pattern of clustering that Charlier devised was the first application in science of what later became known as a 'fractal' distribution, after Benoît Mandelbrot introduced this term in 1972 to describe patterns which repeatedly copied themselves on larger and larger scales. This design can be found all over the natural world and it provides a simple way of understanding the branching of trees or the pattern of the human metabolic system. Natural selection has favoured fractal clustering in any situation where it is advantageous to develop

Figure 4.1 Carl Charlier (1862–1934).

large surface area (to absorb nutrients, for instance) with a minimum of volume and weight (Figure 4.3).

Charlier's universe was really an extension of the Copernican principle that Einstein had used to simplify models of the universe: it presents the same clustering pattern on any scale that you view it. His model was carefully constructed. Despite being infinite in volume its clustered matter extended without limit and was not confined within a finite island. The clustering died away fast enough for the integrated light from all the stars not to make the sky always bright,[6] the gravitational force acting at every place in the universe was similarly finite, and the speeds at which stars were moved by gravity remained small on all scales of clustering.[7]

This type of cosmological scenario, which side-stepped the problems of using Newton's theory in an infinite universe and overcame Einstein's objections to a Newtonian description of space, was taken

Figure 4.2 A fractal distribution displaying three levels of hierarchical clustering. Each block represents a 'galaxy' which is clustered into larger 'clusters' of eight blocks that are in turn clustered into a single 'supercluster' block. This process can be continued upwards and downwards in scale without limit.

up energetically in 1922 by a Viennese philosopher and self-educated physicist, Franz Selety (1893–1933).[8] In 1922, Selety published a clear expression of the hierarchical universe picture in the leading physics journal of the day showing how all Einstein's objections to infinite Newtonian universes could be met.[9] Simply put, he provided a recipe for an infinitely large hierarchical universe which contained an infinite mass of clustered stars filling the whole of space, yet with a zero average density[10] and no special centre.

The problem faced by proponents of this type of universe was explaining how matter came to be distributed in this 'nice' nested hierarchy of clusters in the first place. Einstein had already raised the concern that even if it could arise it would dissolve away into another structure as stars randomly escaped from clusters and were captured by the gravitational pull of nearby ones. Selety had to admit that it was extraordinarily improbable that such a structure would arise in

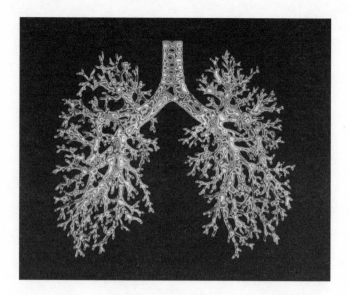

Figure 4.3 The fractal distribution of the bronchial system in the human lungs. The trachea branches into smaller and smaller tubes, or bronchi, so as to maximise the exposed surface to pass air with a small total volume.

the first place, but so long as the density fell off like the inverse square of the distance as you moved up the hierarchy of clusters, the random speeds of the stars would not be large enough for the whole hierarchy to drift apart in a finite future time.[11]

Einstein responded[12] quickly to Selety's article, noting that it overcame the difficulties that he had wrongly assumed to be inevitable for Newtonian cosmology and admitting that a 'hierarchic construction of the universe is *possible* . . . but unsatisfactory'.[13] Selety continued to push his cosmology in articles published[14] in 1923–4 and others took it up as well, notably the eminent French mathematician Émile Borel, but Einstein never commented on it again.

The celebrity of the hierarchical cosmology was tantalisingly brief, yet in retrospect Selety was entirely correct. It occasionally re-emerged in the context of Einstein's theory in the 1970s,[15] and again in the 1990s, in attempts to accommodate observations of galaxy clustering. Today, the detailed observations of the temperature fluctuations in the background radiation indicate that the clustering does not continue indefinitely in a fractal structure.[16]

DR. KASNER'S UNIVERSE

I hope I never make the mistake of not making the right
mistakes.
    Samir Samaje

You may well have heard of a thing called the World Wide Web, and
no doubt also of the remarkable search 'engine' called Google which
has wrapped its tentacles around the information content of the world
and can narrow down a search for a piece of replacement crockery
or a book with surprising swiftness. 'Google' is a curious name for
the company that runs this huge computer-intensive search operation;
its headquarters, the Googleplex, sounds even odder.

The story behind those names begins with an American mathematician, Edward Kasner, who was a professor at Barnard College,
Columbia University, in New York (Figure 4.4). In addition to his
research work in different areas of mathematics, he was also passionate
about communicating mathematics to the general public and to young
people through talks, books and articles. His most famous publication
was the book *Mathematics and the Imagination*, which he wrote with
James Newman, first published in 1940 and still in print today. One
of its chapters dealt with very large numbers and gives an example
of a neat-looking number that is enormously large: 10^{100}; that is, 1
followed by 100 zeros. (For comparison, there are only about 10^{80}
atoms and 10^{90} photons of light in the entire visible universe). In 1938,
Kasner's nine-year-old nephew, Milton Sirotta, invented the name
Googol for this number and then coined the term *Googolplex* for the
unimaginably larger number obtained by raising the googol to the
power of ten, so:

$$1 \text{ Googolplex} = 10^{Googol}$$

This number is so large that if I started to write it out in full as
10000000000. . . it would not fit into the visible universe, which is
only about 10^{29} cm across.

The story told by the computer scientist David Koller[17] is that back
in 1996 two young computer science Ph.D. students at Stanford

Figure 4.4 Edward Kasner (1878–1955).

University, Larry Page and Sergey Brin, were starting to think about how they could map the network of interconnections between different pages containing common words and citations on the Web. Eventually, the strategy they developed to rank pages turned into the most effective internet search engine. At first they called this new search technology 'BackRub', but the following year they were trying to find a better name to reflect the huge number of links involved in the searches, and one of their fellow students, Sean Anderson, suddenly suggested Googolplex, which Page shortened to Googol. This sounded like a good name so Anderson did a quick computer search to see if Googol.com was still available as a possible internet domain name. In his rush he miskeyed the spelling as 'Google.com' and found that it was still available. Brin seemed to like the new (mis)spelling better than the original and registered Google.com under Brin and Page's names that very day, 15 September 1997. Later, when Google grew into a massive company, its remarkable headquarters building in

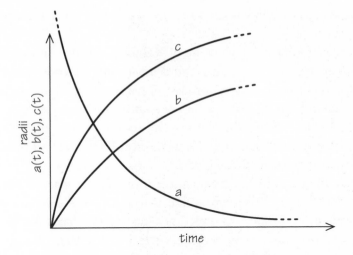

Figure 4.5 The expansion of the three radii, X, Y and Z, of the Kasner universe. Two directions expand but the third contracts. The total volume, XYZ, always grows in direct proportion to the time.

Mountain View, Santa Clara, close to San Jose in California, was nicknamed 'The Googleplex'.

Edward Kasner did more than propagate some new words for big numbers to the world at large. In 1921, he was among that very small group of people trying to find new solutions to Einstein's equations. He knew of the early solutions of Einstein and de Sitter which included the repulsive cosmological constant. Unlike these famous forerunners, however, Kasner had no special knowledge of astronomy. He was well versed in the abstract mathematics behind Einstein's equations and set himself the mathematical challenge of finding solutions in the situation where the matter in the universe could be neglected. This was an assumption that de Sitter had made but, unlike de Sitter, Kasner dropped the cosmological constant from the equations completely. To make up for these simplifications Kasner introduced a completely new possibility: that the universe could expand at different rates in different directions. Kasner's anisotropic universe had space with flat Euclidean geometry. It was infinite in size, and expanded from a beginning at a finite time in the past and continued to expand for ever. Whereas parts of the expanding universes of Friedmann and Lemaître could be visualised as spheres which expanded in size, Kasner's universe was like an ellipsoid that expanded in volume growing at different rates in every direction.

There was one very striking feature of Kasner's universe. Although it expanded in volume, it actually *contracted* in one direction while expanding in the other two directions at right angles to it (Figure 4.5). Hence a sphere of empty space in Kasner's universe would get larger around its Equator, which would become increasingly elliptical, while the two poles would contract towards the centre. The universe would become increasingly shaped like a pancake.[18]

This is a strange universe. It contains no matter. Its space is not curved. Yet it expands. It is a world that is driven by the differences between expansion in different directions. We are familiar with the effects of variations in gravity because we feel their effects through the tides. The Moon (and the Sun) pull on the oceans more strongly on the side of the Earth that is closer to the Moon and so the sea level is higher on that side of the Earth than the other. The tidal influences change steadily because of the daily rotation of the Earth and vary with the inverse cube of distance rather than the inverse square of Newton's gravitational force between masses. Kasner's universe contains this tidal gravitational force and it is different in each direction. It allows a universe to be 'started' by expanding at different rates in different directions.[19]

Kasner's universe looks very special. If any matter is added to it then it will gradually evolve towards the isotropic universe of Einstein and de Sitter[20]. If the repulsive cosmological constant is added to it then it will expand towards the exponentially expanding de Sitter universe, regardless of whether there is also any matter in it. However, if we follow it backwards in time to its 'beginning' at $t = 0$ then neither matter, radiation nor the cosmological constant have any significant influence upon how it looks. The expansion rate differences win out and the universe begins expanding from an infinitely stretched 'needle' of zero volume, with infinite radius in one direction and zero radius in the two directions at right angles to it, rather than a 'point'. Despite its special nature, the Kasner universe will turn out to have huge significance for our understanding of possible universes.

DIRAC'S UNIVERSE — WHERE GRAVITY DECAYS

I don't see how you can work on physics and write poetry at the same time. In science, you want to say something nobody knew before, in words everyone can understand. In poetry, you are bound to say something that everybody knows already in words that nobody can understand.

Paul Dirac[21]

Paul Dirac (1902–84) was the Lucasian Professor of Mathematics at Cambridge for some of the time when Eddington was living and working at the University Observatories. He is arguably the greatest British physicist of the twentieth century, responsible for creating much of the theory of quantum mechanics and the mathematical language in which it is conventionally expressed, predicting the existence of antimatter, discovering the statistical properties of populations of elementary particles and the key 'Dirac equation' that describes the behaviour of relativistic electrons. He was the youngest winner of the Nobel Prize for physics in 1933, when just thirty-one years of age, having been appointed to the Lucasian professorship a year earlier.

Dirac's biographer[22] described him as 'the strangest man' and his Cambridge contemporaries could hardly have been said to 'know' him. He was a man of few words who summed up others very succinctly: Ludwig Wittgenstein, for example, he described as an 'awful fellow, never stopped talking'.[23] Stories of Dirac's simple and entirely logical approach to life and awkward social behaviour are legion and it is not surprising to find that his unexpected foray into cosmology was sent to the journal *Nature* a few days after he returned from his honeymoon, in February 1937. It was followed ten months later by a much longer article proposing a new basis for cosmology. He didn't write about it again for another thirty-five years but then took it up again as though no time had passed.

Most physicists of the time, when told that Dirac had written a research paper on cosmology, would have been expecting a comprehensive new theory or a new way of solving Einstein's challenging equations using methods of great mathematical sophistication. Instead,

Figure 4.6 Paul Dirac (1902–84).

Dirac's was a fascinating but very simple idea that many felt to be eccentric – 'Look what happens to people when they get married,' quipped his friend Niels Bohr. Dirac argued that if we encounter in physics very large dimensionless numbers taking values like 10^{40} or 10^{80}, then they are most unlikely to be independent and unrelated. Most probably there is an undiscovered mathematical law of Nature linking the quantities involved. This is Dirac's Large Numbers Hypothesis (LNH).[24] The large numbers that Dirac had noticed were threefold, and involved the age of the universe t, the speed of light c, the masses of an electron m_e and a proton m_p, and Newton's gravitation constant G. From them he constructs the following three numbers:

N_1 = the ratio of the size of the observable universe to the electron radius

$$= ct/(e^2/m_e c^2) \approx 10^{40}$$

N_2 = the ratio of the electromagnetic and gravitational forces between a proton and an electron

$$= e^2/Gm_e m_p) \approx 10^{40}$$

N = number of protons in the observable universe

$$= c^3 t / Gm_p \approx 10^{80}$$

By his hypothesis,[25] the numbers N_1, N_2 and $\sqrt{N}$ were likely to be *equal*, to a very good approximation. It would be very, very odd if such enormous numbers had no connection at all. Dirac believed that there must be unknown laws of Nature that require formulae exactly (or approximately) of the form $N_1 \approx N_2$ or $N \approx N_2^2$.

The radical change precipitated by Dirac's LNH is that it requires us to believe that *a collection of traditional constants of Nature must be changing as the universe ages in time,* t, because he requires

$$N_1 \approx N_2 \approx \sqrt{N} \propto t$$

As a result, a combination of three of the traditional constants of Nature is not constant at all, but must increase steadily in value as the universe ages, and t increases, so:

$$e^2 / Gm_p \propto t \quad (\star)$$

Dirac chose to accommodate this requirement by abandoning the constancy of Newton's gravitation constant, G. He proposed that it was decreasing in direct proportion to the age of the universe over cosmic timescales, falling as

$$G \propto 1/t$$

and this satisfies the requirement of equation($\star$). So, in the past G was bigger, and in the future it will be smaller, than it is measured to be today. One now sees that $N_1 \propto N_2 \propto \sqrt{N} \propto t$ and the huge magnitude of all three Large Numbers is a consequence of the great age of the universe:[26] they all get larger as time passes.[27]

Dirac's approach had three significant elements. First, he sought to show that what might previously have been regarded as coincidences are actually *consequences* of a deeper set of relationships that had been missed. Second, it led him to conclude that the curvature and the cosmological constant in the universe must be zero; otherwise they would give rise to further large numbers. Third, he sacrificed the constancy of the oldest known constant of Nature. Unfortunately, Dirac's hypothesis did not survive for long. The proposed change in the value of G was just too dramatic. Dirac suggested that in the past gravity would have been much stronger. The energy output of the Sun would be changed and the Earth would have been far hotter in the past than usually assumed[28]. As the American physicist Edward Teller argued in 1948, the oceans would have been boiling in the pre-Cambrian era and life as we know it would not have evolved.[29]

Teller's friend George Gamow responded to the boiling-ocean problem by suggesting that it could be ameliorated if it was assumed that Dirac's coincidences were created by a time variation in e, the electron charge, with e^2 increasing with time as the equation (*) requires.[30]

This suggestion didn't survive for long either. Unfortunately, Gamow's proposal for varying e had all sorts of unacceptable consequences for past life on Earth. It was soon realised that his theory would also have resulted in the Sun exhausting all its nuclear fuel long ago. The Sun would not be shining today if e^2 grows in proportion to the age of the universe.

Despite its failure, Dirac's suggestion opened up new possibilities for our models of the universe. Changing the constant of gravitation into a time-varying quantity amounted to proposing that Einstein's general relativity theory was incorrect, or incomplete. You can't just turn a constant into a variable in the cavalier way that Dirac proposed. In Einstein's picture of gravity, all forms of energy gravitate. They contribute to curving space and determining the rate of flow of time. Changes in G need to contribute too. Others now sought to put the whole idea on a firm foundation by creating a slight variant of Einstein's theory which accounted for the varying 'constant' of gravitation as though it were a new source of energy and gravitation.

The first to develop a full theory like this, in 1939, by extending Einstein's relativity theory to contain a new energy field which represented the variations in the traditional gravitation constant, G, was the infamous German physicist Pascual Jordan. Jordan had made important contributions to quantum mechanics in a series of papers with Max Born and Werner Heisenberg. Unfortunately, Jordan was also a keen Nazi who joined the National Socialist party in 1933, the year in which Hitler took power, moving on to become a member of the brownshirt stormtroopers in 1934.[31] He joined the Luftwaffe in 1939 and worked on meteorology during the Second World War, spending a period at the Peenemünde V-1 and V-2 rocket research facility on the Baltic coast. Yet, despite his enthusiasm to work on weapons systems, he does not appear to have been trusted fully by his superiors, possibly because of his past association with Jewish physicists, like Born. Jordan's political sympathies created a great gulf between him and other physicists until the early 1950s, when he was restored to an academic post. Hence his work in cosmology during the previous decade was largely ignored at the time. Some have said that his political activities cost him a share in the 1954 Nobel Prize for physics.

Although Dirac's excursion into cosmology was short-lived and superseded by more precise formulations like Jordan's, the numerical coincidences that Dirac noticed turned out to have an interesting interpretation. Robert Dicke pointed out that the formula $N \approx N_1^2$ is just the statement that the time, t, at which we are observing the universe (which we call its 'age') is roughly equal to the time it takes for a star to form and settle down into a stable period of evolution, burning hydrogen into helium by nuclear fusion reactions. Since we couldn't exist before stars form, and are unlikely to exist after they die, this is not such a surprising coincidence after all.[32] We probably could not exist in a universe where we didn't see the 'coincidence' (*).

Dirac had not realised that we are observing the universe at a special period in its history. There are only certain intervals of time when life of any sort is possible in an expanding universe and we can practise astronomy only during that habitable time interval in cosmic history.

Dirac responded to this important objection of Dicke's by agreeing that there was a time before which life could not exist in an expanding universe, because it would be too hot and dense; but he believed that, once begun, life could continue in the universe for ever.[33] For Dirac, life wasn't confined to a short interval of cosmic history: his faith was that its future was unlimited.

EINSTEIN AND ROSEN'S UNDULATING UNIVERSE

> . . . small portions of space are in fact analogous to little hills on a surface which is on the average flat, namely that the ordinary laws of geometry are not valid in them. That this property of being curved or distorted is continually being passed on from one portion of space to another after the manner of a wave.
>   William Clifford[34]

By 1932, Einstein had left Europe for the newly created Institute for Advanced Study in Princeton, New Jersey, after a short stay in Oxford as a research lecturer at Christ Church College. Far away from the troubled political situation in central Europe, he started to think about solutions to his equations again and was pleased to have a young research assistant, Nathan Rosen, come to help him with his mathematics, in 1935. Rosen would co-author several of the most famous papers in theoretical physics with Einstein during the next two years.[35] By 1936, Einstein and Rosen had found a new type of solution to Einstein's equations. It described a universe that expanded but had the symmetry of a cylinder, so that everything depended on time and one of the directions in space. This symmetry simplifies the fantastic complexity of Einstein's equations and allows an exact solution to be found. This universe has a dramatic new property not seen in any of the other possible universes that had been found from Einstein's equations. It contains waves which travel through space, causing the geometry of space to ripple as they go. It is as if Kasner's universe, expanding at different rates in different directions, has had some waves added to it which travel outwards from a line which is an axis of symmetry, like a tube of kitchen roll that is throwing off paper outwards as it unwinds (Figure 4.7).

The most interesting thing about this universe for Einstein and his contemporaries was the presence of these 'gravitational waves'. The Einstein–Rosen universe contained no matter, so any waves would be ripples in the geometry of space travelling through time (Figure 4.7). The idea of gravitational waves had been around for some time and was controversial. Some people believed that they were just 'paper

Figure 4.7 The Einstein–Rosen universe contains cylindrical gravitational waves which propagate outwards from a line running through an expanding universe.

waves' which were consequences of a particular set of co-ordinates for expressing Einstein's equations and didn't correspond to any real waviness in space. Others believed they were physically real. If one came your way it would have an effect on you (stretching you in one direction and compressing you in the perpendicular direction like a tidal force).

Einstein and Rosen soon realised that their new solution provided a perfect testing ground in which to settle this dispute without recourse to uncertain approximations or numerical calculations. Remarkably, the conclusion they first came to was that the cylindrical gravitational waves were not 'real': they were just artefacts of their choice of co-ordinates. They were saying it was just like looking at a geographer's globe and seeing that the meridians all intersect at the North and South Poles. To the uninitiated this might suggest that something bad happens to the Earth's surface there, with a great convergence of things, but nothing does happen. You can always switch to some other map co-ordinates at the Poles and everything will be unremarkable. Imagin-e that you had chosen wiggly lines to represent the lines of

longitude. Again, to the uninitiated those lines might suggest something undulatory going on at the Earth's surface – yet you would be
wrong. However, beware, because if you looked at a contour map
you would find complicated undulatory lines that might look rather
similar. In this case you would be wrong to conclude that they corresponded to nothing 'real' on the Earth's surface. Einstein and Rosen
were faced with trying to decide whether their waves of curvature
were 'real' like the contour lines or purely human constructions like
the wiggly lines of longitude.

In early summer in 1936 they sent their paper off to the leading
American physics journal of the day, the *Physical Review*. They claimed
that the gravitational waves in their new mathematical universe were
not physically real. After their paper was received by the journal on 1
June it was handled like all other submissions and sent off to another
scientist to evaluate. On 23 July, the report on the paper came back
and was sent to Einstein without the identity of the reviewer being
revealed to him (as is usual in this process). We know now that the
scientist reporting to the editor was none other than Howard
Robertson, one of just a handful of American scientists familiar with
the technical details of general relativity. He was unconvinced by the
conclusions of the paper and pinpointed the place where the authors
had drawn a wrong conclusion from the evidence. He clearly thought
the gravitational waves in their universe were real and asked the
authors to consider his comments about this. Before hearing Einstein's
response it is important to remember that in Europe, where all
Einstein's early work had been published, it was not the convention
for papers to be sent out for peer review in this way. They would be
published as they stood if the author was of good reputation or if
they were communicated to the journal by an established scientist on
behalf of the author, or they would be evaluated by the editor alone.
To reject a paper submitted by an established author would have been
rather insulting and it rarely happened. Consequently, not understanding the American system, Einstein was upset to hear that the
editor had sent his paper to another scientist. He replied to the *Physical
Review*'s editor, John Tate:

We (Mr Rosen and I) had sent you our manuscript for *publication* and had
not authorized you to show it to specialists before it is printed. I see no

reason to address the – in any case erroneous – comments of your anonymous expert. On the basis of this incident I prefer to publish the paper elsewhere.[36]

Einstein quickly submitted the paper to the *Journal of the Franklin Institute*, where he had published a paper in the past.

Over the next few months he remained convinced that his new universe contained no real gravitational waves. All this changed after Einstein developed a friendship with Robertson – the same Howard Robertson who unbeknown to him had refereed their paper for the *Physical Review* and cast doubt upon their conclusions! Robertson persuaded Einstein that there was a better way to represent the solution that he and Rosen had found and there was no doubt that the cylindrical gravitational waves were real. By now, the Einstein–Rosen paper had been accepted for publication in the *Journal of the Franklin Institute*, but Einstein was able to make a face-saving change to the conclusions when the proof pages came for him to check, adding a note of thanks to 'my colleague Professor Robertson for his friendly assistance in the clarification of the original error'.[37]

Nathan Rosen had left for the Soviet Union to work at the University of Kiev just after their paper had been submitted and only learned of all these developments when he read in the newspaper about the appearance of a new Einstein paper, which was a newsworthy event. Rosen was not persuaded by Einstein and Robertson's arguments for the reality of the gravitational waves and subsequently published his own paper, still insisting upon the earlier erroneous conclusion. He continued arguing against them until the 1970s, by which time almost every other physicist had long been convinced of the reality of gravitational waves. A crucial argument was a simple point made by Richard Feynman at a conference in Chapel Hill, North Carolina, in 1957, two years after Einstein's death. He proved that a bead placed upon a rough rod (a so called 'sticky bead' because of the friction) would move backwards and forwards when gravitational waves passed at right angles to the rod. The roughness of the rod would lead to frictional heating when the bead moved, just like when you rub your hands together to warm them up. This warming meant that the gravitational wave was the source of the heat and so must carry energy. Hence gravitational waves didn't just exist on paper.

Feynman's 'sticky bead' argument settled the issue for almost everyone who had doubted the reality of gravitational waves. They were like tidal forces. If a gravitational wave passed through this page, face-on, then it would stretch the page in the one direction and squeeze it in the direction at right angles to it: a square would turn into a rectangle and a circle into an ellipse.

There is an amusing postscript to this story. Feynman registered under a pseudonym at the Chapel Hill conference because he seems to have had a low opinion of its agenda and the 'sticky bead' argument was therefore posted anonymously. In his memoirs, *Surely You're Joking, Mr Feynman!*, he tells of his efforts to find the conference venue:

One time, in 1957, I went to a gravity conference at the University of North Carolina. I was supposed to be an expert in a different field who looks at gravity. I landed at the airport a day late for the conference (I couldn't make it the first day), and I went out to where the taxis were. I said to the dispatcher, 'I'd like to go to the University of North Carolina.'

'Which do you mean,' he said, 'the State University of North Carolina at Raleigh, or the University of North Carolina at Chapel Hill?'

Needless to say, I hadn't the slightest idea. 'Where are they?' I asked, figuring that one must be near the other.

'One's north of here, and the other is south of here, about the same distance.'

I had nothing with me that showed which one it was, and there was nobody else going to the conference a day late like I was.

That gave me an idea. 'Listen,' I said to the dispatcher. 'The main meeting began yesterday, so there were a whole lot of guys going to the meeting who must have come through here yesterday. Let me describe them to you: They would have their heads kind of in the air, and they would be talking to each other, not paying attention to where they were going, saying things to each other, like "G-mu-nu. G-mu-nu."'

His face lit up. 'Ah, yes,' he said. 'You mean Chapel Hill!' He called the next taxi waiting in line. 'Take this man to the university at Chapel Hill.'

'Thank you,' I said, and I went to the conference.[38]

5 Something Completely Different

I once found the philosopher Richard Rorty, standing in a bit of a daze in Davidson's food market. He told me in hushed tones that he'd just seen Gödel in the frozen food aisle.

Rebecca Goldstein[1]

A SWISS-CHEESE UNIVERSE

I must admit to being disappointed to read that Einstein's telephone number was 2807.

Michael Mahler

The 1940s brought the study of universes to a grinding halt. The spread of war led to the redeployment of physicists and mathematicians into weapons research, meteorology, aeronautics and cryptography. Universities were closed to the arrival of new students and international contacts were restricted and limited to close allies. Einstein was in the United States and many other great German scientists had fled to Britain and America. The universe had never seemed smaller.

By 1944, Einstein had recruited a new assistant at Princeton. His assistants were always talented young mathematicians who could make up for Einstein's self-confessed weakness in this area. Ernst Straus was something of a mathematical prodigy. He started taking interesting shortcuts in calculating sequences of numbers when only five years old, spotting the trick that enabled you to add up the first 100 numbers in your head in a few seconds.[2] He was born in Munich in 1922 but after the Nazis came to power in 1933 his family fled to Palestine, where he was educated at high school and at the Hebrew University

Figure 5.1 Ernst Straus (1922–83).

in Jerusalem. Straus didn't stay to take an undergraduate degree and instead, while still a teenager, moved to New York's Columbia University in 1941 to begin graduate research. In 1944, he found himself recruited as Einstein's new research assistant at the Institute for Advanced Study in Princeton.[3]

The young Straus had no background in physics and his mathematical inclinations were towards number theory and 'pure' mathematical topics but he lost no time in filling the gap left by the departures of Nathan Rosen (1935–45) and Leopold Infeld (1936–8). By the spring of 1945, Einstein and Straus had found a new type of possible universe using Einstein's equations.[4] It described a universe which looked largely like one of the simple expanding universes of Friedmann and Lemaître containing material (like galaxies) which exerted no pressure. But it had spherical regions removed from it, like bubbles in a Swiss cheese (Figure 5.2). Each empty hole then had a mass placed at its centre. The mass was equal in magnitude to what had been excavated to create the hole. This was a step towards a more realistic

a
b

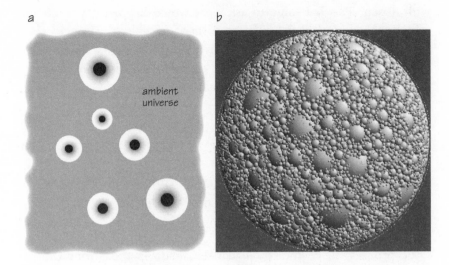

ambient
universe

Figure 5.2 (a) The Swiss-cheese universe model. Spherical regions are evacuated, leaving all their mass in a spherical lump at their respective centres. (b) A computer-generated Swiss-cheese universe created by Allen Attard of the University of Toronto. It begins with a uniform density, then 34,667 spheres of different sizes are marked out and all the matter in each sphere is compressed to a point at their respective centres; 80 per cent of the space is contained in the spherical holes.

universe in which the matter was not smoothly spread with the same density everywhere, but gathered up into lumps, like galaxies, which were spread about in empty space.

Each of the 'holes' was spherical and this new Swiss-cheese universe could be accommodated in Tolman's non-uniform universe by a suitable choice of starting conditions. As always, the discovery of an exact solution to a collection of equations as complicated and difficult as Einstein's meant that there had to be some simplifying feature in the solution that made the equations tractable. Yet this is reminiscent of Groucho Marx's famous claim that he did not want to belong to any club that would have him as a member:[5] any solution of Einstein's equations that is simple enough to find will invariably contain a special feature that might render it atypical or uninteresting.

Einstein and Straus's solution was simple because it was spherical and so excluded the possibility of any gravitational waves being present – unlike the case of the cylindrical universe that Einstein had found with Rosen. This led some to ask what would happen if you could somehow combine all the different sorts of irregularity at once. The presence of all these irregular features would of course crush any hope of solving Einstein's equations. Yet there was one way to get a glimpse of what such a universe would be like.

PERTURBED UNIVERSES

'There are some trees, Watson, which grow to a certain height, and then develop some unsightly eccentricity.'
    Arthur Conan Doyle[6]

Little progress in cosmology had been made in the Soviet Union following Friedmann's remarkable discovery of the expanding universe in the early 1920s. Einstein's theory was regarded as inextricably linked to questionable idealistic philosophies which were contrary to the dictates of dialectical materialism.[7] And the evils of Stalin's dictatorship gave everyone enough to worry about. The greatest theoretical physicist in the country, Lev Landau, was arrested and imprisoned during the Great Purge, for anti-Soviet activity in 1938–9. He was released only because of the persuasive intervention of internationally renowned physicists like Peter Kapitsa and Niels Bohr.

Landau had become professor at the new Ukrainian Physicotechnical Institute in Kharkov in 1933, when only twenty-seven years of age. He had been making outstanding contributions to physics since he had been a teenager and would eventually win the Nobel Prize in 1962 for his explanation of the superfluid behaviour of helium at low temperatures.[8] At the new university, Landau's reputation attracted outstanding students and he set new standards for advanced physics education. His best-known student was a young man from Kharkov named Eugene Lifshitz. Lifshitz finished his degree in physics and mechanics when he was eighteen and took little more than a year to master Landau's challenging course and pass all his examinations for his Ph.D. degree.[9] Years later, Lifshitz would co-author with Landau a series of advanced texts, known to physicists everywhere simply as 'Landau and Lifshitz', which set out the core of theoretical physics needed for research in an elegant, economic and unified style.

Landau's arrest was potentially a disaster for Lifshitz because of his close association with his mentor.[10] Fortunately, the climate improved when Landau was released from prison and he and Lifshitz moved to the Institute of Physics Problems in Moscow. But there any pattern of normal work was soon broken by the outbreak of the war. The Battle of Moscow began in September 1941 and lasted until January 1942. With the city facing collapse and starvation, many people fled to the east or were evacuated. Many of the physicists were moved to Kazan. Moscow was saved by the elements: very heavy rain and bitterly cold weather, for which the invading German troops were inadequately prepared, halted the advance. It was in this period of hardship and deprivation that Lifshitz began to study a new cosmological problem. Like Friedmann before him, his motivation was mathematical, not astronomical.

We have seen that scientists had become interested in what happens in universes which are not isotropic and uniform in space, like the ones found by Friedmann, Lemaître and de Sitter. First, Kasner and Tolman, and then Einstein, Rosen and Straus, had started to explore the properties of universes which differed from one direction to another and from place to place. These are clearly more realistic because the real universe is not perfectly smooth and identical in every direction.

Lifshitz approached the problem in a way that is very common in physics: you take the simple exact solution of your theory and then see what happens if you change it by a very small amount. Do the equations require these little changes to die away as time increases or do they grow and amplify? If the little changes die away it tells us that the simple universe we decided to perturb is stable and doesn't evolve into one with a very different structure. However, if the small changes grow with the passage of time our simple universe is unstable and will evolve into a different state in the future.

In his paper of 1946, published in Russian, Lifshitz examined what happens if you slightly perturb the isotropic and homogeneous universes of Friedmann.[11] He seems to have treated this as a purely mathematical problem and did not relate it to the question of why irregularities like galaxies exist in the universe. He showed how three distinct types of small irregularity can be created. The first was simply to vary the density of matter a little from place to place; the second was to spin the matter slowly; and the third was to introduce a small ripple of gravitational waves into the smooth space. When their effects are very small these different effects don't get mixed up into a turbulent muddle like they do when they are large. Their effects remain separate and simple to extract. If we took a spherical ball and applied these three perturbations to it then the first would correspond to varying its density and shape, the second to spinning it, and the third to squashing it like an ellipsoid but without changing its volume.

Lifshitz showed that small differences in the density tend to become more pronounced as time passes. This was something that Newton had also understood at the end of the seventeenth century, when he pointed out in a letter to Richard Bentley, long-time (tyrannical) Master of Trinity College, Cambridge, that if a perfectly smooth distribution of matter were made slightly irregular, then the over-dense regions would attract more material towards them at the expense of the sparser ones and things would get increasingly irregular. We call this 'gravitational instability' (Figure 5.3). It happens fairly quickly in a distribution of material in a space that is not expanding. Lifshitz showed that it still happens in an expanding universe but the irregularity grows more slowly because the clumping together of different particles of matter has to work against the opposing effect of the expansion pulling them apart. He also showed that, if you look at small enough clumps, the process of increasing irregularity eventually gets stopped if there is any pressure to oppose it. The overall picture was simple. If you start off with a universe that is *almost* isotropic and homogeneous then the deviations from uniformity get more and more pronounced as the universe expands.[12] This was the first calculation to show that isotropic and homogeneous universes are special. The fact that after billions of years of expansion all the non-uniformities in the universe are still rather small on the average implies that they must have started out fantastically small 14 billion years ago.

Gradually, other astronomers tried to use these calculations to arrive at an explanation for the existence of structures such as galaxies in the universe. Unfortunately, these explanations were not totally compelling. In order to get the irregularities to increase to form galaxies in the recent past you had to choose a special level of irregularity that existed at some time close to the beginning of the expansion. Yet as

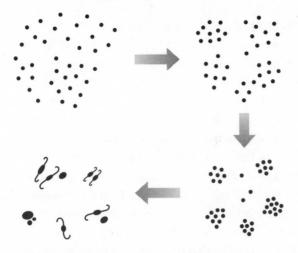

Figure 5.3 Gravitation gradually transforms a slightly non-uniform distribution of material into an increasingly clumpy one. The denser regions exert more gravitational pull and accrete additional material at the expense of sparser ones.

there was no good reason for such irregularity to exist at any special level at any time, this was not really an explanation of anything.

When it came to the other types of irregularity in Lifshitz's calculations, one could show that any large spinning vortices of matter will spin more slowly as the universe expands. Later, some cosmologists would take this to mean that they must have been huge in the distant past, which suggested to them that all the rotating galaxies we see today must have formed out of a chaotic primordial turbulence. Lastly, the gravitational waves that Einstein and Rosen had struggled to interpret correctly also appeared as a possible small disturbance to a perfectly smooth universe. Lifshitz's calculations showed that they diminished in strength as the universe expanded and stretched out the waves.

Lifshitz's calculations, carried out as a mathematical exercise, were a landmark in the study of universes. They have been repeated since then in dozens of different ways and the results used to determine the signatures of all three types of small perturbation when we observe the background radiation or the clustering of galaxies. Whole books have been devoted to their consequences. They provide us with a way of understanding how irregularities develop in the universe if they start from small beginnings.

SCHRÖDINGER'S UNIVERSE

Everything in the future is a wave, everything in the past is a particle.
        Lawrence Bragg[13]

While Einstein revolutionised our understanding of the universe at large, there occurred an even more rapid transformation of our understanding of matter on the atomic scale. The emergence of quantum mechanics as a new way of making sense of the behaviour of matter and light owed a lot to Einstein, but more to Niels Bohr, Werner Heisenberg, Paul Dirac, Max Born, Wolfgang Pauli and Erwin Schrödinger. It steadily revealed the structure of atoms and molecules, explained the Periodic Table of the chemical elements and predicted a host of the properties of solid materials. Yet it was a subject that had no intersection with cosmology. Dirac had dabbled in cosmology but he didn't make any use of quantum mechanics in his claims that gravity might be weakening in strength as the universe aged. Lemaître had an idea of how the universe emerged from some type of superdense initial state that he dubbed the 'primeval atom'. To understand it, quantum mechanics would need to be joined to Einstein's theory of gravity but Lemaître didn't know how that could be done. Some physicists wondered if the creation of matter might be described by the new quantum mechanics, and Richard Tolman speculated vaguely that the centres of galaxies might be the sources of an ongoing interconversion of matter and radiation. None of these speculations led anywhere, but then one notable physicist managed to provide a first insight into the idea of a quantum universe.

The equation Erwin Schrödinger found in 1926 and which bears his name is the most important equation in all of mathematical physics. Its solutions describe all atomic and molecular structure, all of material science and chemistry. Schrödinger was the only son of well-to-do parents who recognised his abilities and provided him with private tutors. He became a successful student at the University of Vienna and then a lecturer there, before holding professorships in Zurich, Berlin, Oxford and Dublin. He would share the Nobel Prize for physics with Paul Dirac in 1933.

Figure 5.4 The grave of Erwin Schrödinger (1887–1961) in the Austrian village of Alpbach bears an inscription of the Schrödinger equation.

In his personal life Schrödinger was rather unorthodox. His short stay at Magdalen College, Oxford, in 1933, and attempts to tempt him to a position in Princeton the following year, were complicated by his desire to set up house with both his wife and his mistress; all three of them were in turn romantically entwined with other physicists and their spouses. In the end he moved to Graz in Austria in 1936 but was invited to Dublin in 1940 by the Irish Taoiseach, Éamon de Valera, to join the new Dublin Institute for Advanced Study, styled on the Princeton body that had been established for Einstein. He accepted but the end of his career in Dublin was marked by a number of minor scandals involving affairs with married women and illegitimate children. If Schrödinger's lifestyle was a little much for Oxford in the 1930s, it was far too much for Ireland in the 1950s. Schrödinger died in 1961 and is buried in the churchyard of the beautiful Tyrolean village of Alpbach, overlooked by the Alpine landscape that he loved. His grave memorial has the Schrödinger equation inscribed upon it (Figure 5.4).

Schrödinger took a keen interest in the expanding universes of

Einstein's theory of general relativity. His first foray into the study of universes was an attempt to see what quantum mechanics might have to say about the universe. He decided to investigate what happens to waves in an expanding universe.[14] The results provided insights not only into the behaviour of ordinary sound waves and light waves but into the spread of quantum waves that tell us the probability of observing particular events.[15]

Schrödinger made an important discovery that no one (including him) fully appreciated at the time. The expansion of the universe could transform its quantum vacuum energy into real, measurable particles. When there is no expansion, particles and antiparticles continually appear and disappear back into radiation. Energy is conserved, and this picture of seething activity describes the quantum vacuum. However, if this vacuum expands fast enough, or is subjected to a gravitational force that varies very dramatically over a short distance, then those particles and antiparticles appearing from the vacuum can experience forces that are so different that they don't annihilate back into radiation. As a result, real detectable particles and antiparticles appear, feeding off the energy driving the fast expansion or the variations in gravity from place to place. This striking phenomenon of particle production in an expanding universe was studied intensively from the mid-1970s after Stephen Hawking's dramatic discovery that it would also occur at the edge of a black hole and eventually result in the loss of the entire mass of the black hole.[16]

At the time Schrödinger didn't think the process was very important in an expanding universe because its effects are immeasurably small today and totally irrelevant to our understanding of the expansion of the universe. However, in the first moments of the expansion, when the rate of expansion was stupendously high and the radiation density would have been nearly 10^{128} times bigger than it is today, the particle production effects that Schrödinger uncovered would be very significant. They might even explain why the universe displays certain properties. In the 1970s, it was claimed that many asymmetries and irregularities in the expansion would be reduced or removed completely by the particle production process because it feeds off the energy differences from one direction to another and from place to place, gradually damping them down.

Yet cosmologists were not ready for quantum theory in 1939. Had physicists been receptive we can only wonder what the outcome might have been. But Einstein was resolutely opposed to quantum theory ('God does not play dice', as he had stated famously) and would have strongly resisted its application to the universe as a whole. The creation of particles, however, would soon become a very controversial story, as we shall see in the next chapter.

GÖDEL'S SPINNING UNIVERSE

The theological world-view is the idea that the world and everything in it has a good and indubitable meaning . . . Since our earthly existence has in itself a very doubtful meaning, it follows directly that it can only be a means toward the goal of another existence. The idea that everything in the world has a meaning is precisely analogous to the principle that everything has a cause on which the whole of science rests.

Kurt Gödel[17]

When Einstein grew older and his work ceased to solve old problems or pose new ones, he liked to tell people that he went to the office 'just to have the privilege of walking home with Kurt Gödel'. Gödel was one of the greatest mathematicians of the twentieth century and the most important logician since Aristotle. Born in Brno in the old Austro-Hungarian Empire in 1906, one year after Einstein had published his theories of special relativity, Brownian motion and the photoelectric effect, Gödel enrolled at the University of Vienna in 1924 to study physics but was soon seduced by mathematics. He impressed his professors and was soon invited to join the legendary discussion group, the Vienna Circle, which met in coffee shops to discuss philosophy, logic and science and included at various times Ludwig Wittgenstein, Bertrand Russell and Karl Popper. Yet Gödel was the odd man out in this group: he was the only member who did not believe that experience was the sole source of knowledge and that mathematical logic was the only tool with which to solve philosophical problems.

Gödel is most famous for his proof of the incompleteness theorem of arithmetic which, to general astonishment among mathematicians and philosophers, showed that logical systems as rich as arithmetic always contain statements whose truth or falsity cannot be established by using the rules of the system alone. This theorem has all sorts of unexpected consequences; for example, that no computer program that does not alter a computer's operating system can detect all programs that do. Hence no anti-virus program can find all the possible viruses on your computer, unless it interferes with and alters the operating system.

Like Einstein, Gödel came to the Institute for Advanced Study at Princeton to escape the growth of fascism in his home country. He first visited the Institute in 1933–4 but didn't return the following year after suffering some form of mental breakdown.[18] Gödel finally left Vienna with his wife in the autumn of 1939 after being assaulted outside the university by some Nazis who probably mistook him for a Jew. Yet he took the longest way round to Princeton imaginable, across Asia on the Trans-Siberian railway, then by boat from Japan to San Francisco, where he arrived in March 1940, and finally by train across America to Princeton.

By 1942, Gödel had become close friends with Einstein. Both had a similar cultural background and strong philosophical interests, quite different to those of the American scientists around them – and they could converse in German.

On several occasions, Gödel promised Einstein to write down some of his ideas about relativity theory but it was not until 1947 that something exciting emerged. Gödel's letters to his mother at that time tell of his absorption in this work and by the summer of 1947 he had discovered something remarkable. To everyone's surprise, he had been working on trying to find a new solution to Einstein's equations. The result stunned even Einstein.

Gödel's universe was a rotating universe (Figure 5.5). It didn't expand and all the matter rotated at a constant rate about an axis pointing along one direction. It contained Einstein's cosmological constant, but with a negative value so that it reinforced the attractive gravitation of matter, which counter-balanced the effects of rotation pushing matter outwards. This in itself was interesting enough but Gödel's universe had another totally unexpected property: it permitted time travel. Gödel showed that there were paths through space and time that were closed loops. Most people, including Einstein, expected that such a thing would be contrary to the other laws of physics and would

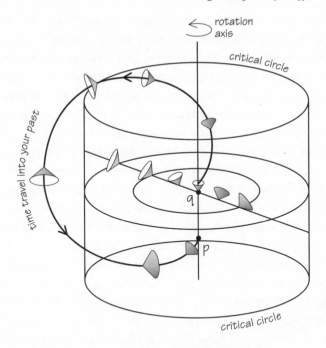

Figure 5.5 Gödel's universe. The material rotates about a central axis with constant angular speed. The rotation affects the light cones which show where light rays travel from each point. As we go farther from the centre the light cones tilt and open out because the linear rotation speed increases. There is a special distance from the rotation centre where the light cones tip over and lie at a tangent to the plane and then straddle it so that light rays can travel below the plane into the past. Suppose your planet was once at p and is now at q. You can visit p again by accelerating to a point outside the critical circle, then travel down to below the original plane through p, then head for the critical circle below p and then go into the future towards p. You have always headed into 'your' future but have ended up in your past.

allow logical contradictions to occur of the kill-yourself-as-a-baby sort that are familiar from science fiction movies.[19] Yet, time travel turns out to be allowed by Einstein's theory and it doesn't conflict with any of the other known laws of Nature. Freeman Dyson remembered hearing about all this in his first meeting with Gödel after he arrived in Princeton as a young man in 1948:

It was in September, 1948. I came to the Institute for Advanced Study as a young member. To my amazement, one of the first people I met was Kurt

Gödel himself, and to my great astonishment he invited me to his home. But anyway, I felt very privileged. He turned out to be very friendly and sociable and not at all the way I'd imagined him. And sane! So he invited me to his home and we talked about physics. Turns out he knew a lot about it and he had actually been working on physics himself, for the previous years he had got a problem from Einstein to look at rotating universes and that's what he did. I was a bit astonished because on the one hand he was an absolutely supreme mathematician, he had done these fantastic pieces of work in mathematics which really shook up the foundations of mathematics . . . and what would such a person do . . . it seemed to be astonishing that he would do something which was so comparatively trivial as prove these rotating universe models existed, and of course it was also not a very interesting part of physics. He himself knew that very well, he was not ignorant of physics, he knew that this wasn't really the mainstream of physics. But anyway, that's the way it was, and so I met him, of course, many more times, and frequently he asked me, 'Have they found it yet? Do they know yet whether the universe was rotating?' He thought that this was something one could really verify by observations and I had to break it to him that observations fell short by about a factor of a million from being able to decide, but he was always disappointed. Every time he called me on the telephone he would usually ask, 'Have they found it yet?' And I always had to tell him no.[20]

In retrospect, Gödel's discovery, which was published as 'An Example of a New Type of Cosmological Solution of Einstein's Field Equations of Gravitation', was of great importance for our study of universes.[21] It revealed that there could be extraordinary global features of the universe that simply did not show up locally. Just because things seemed ordinary in your solar system didn't mean that space and time couldn't get knotted up in strange ways over the scale of the entire universe. Although Gödel's universe wasn't expanding like the universe we see, the unusual properties of time that it displays could still be present in other rotating universes that do expand and resemble our universe.

At first, several notable physicists disagreed that Gödel's universe allowed time travel but they had failed to understand the character of those time-travelling histories and eventually Gödel's deductions were confirmed. In some of his writings he seems to find the possibility of avoiding death by exploiting the circularity of time rather appealing

Figure 5.6 Einstein and Gödel at Princeton.

– a colleague once found him acting it out by writing backwards on his blackboard. Unfortunately, time travel in Gödel's universe requires speeds close to that of light and extremely unnatural configurations of matter. It doesn't look like a very practical proposition for space travellers. We also need to remember that Gödel's universe isn't a manifestation of *Back to the Future*. You can't *change* the past. As Samuel Butler once remarked, even God cannot change the past – only historians can do that.

Suppose that I memorise Shakespeare's *Macbeth* and travel back into the past to meet Shakespeare as a young man, before he has written any of his plays. I tell him in great detail the text and story of *Macbeth*. Shakespeare remembers every word, writes it all down, and publishes *Macbeth*. Where did *Macbeth* come from? I learned it from Shakespeare and he learned it from me. It had no beginning. It just exists.

Logical paradoxes of the 'what-if-I-killed-my-grandmother' variety constitute a genre called 'Grandmother paradoxes' by philosophers interested in time travel.[22] They appear to beset any form of backward-in-time travel.[23] It has been a prominent component of science fiction stories about time travel ever since H. G. Wells's classic, *The Time Machine*, created the idea in 1895.

Should we be persuaded that these 'changing-the-past' examples reveal something fundamentally impossible about time travel? No. There is something not quite coherent about *changing* the past. The past was what it was. You cannot alter it and expect your experienced present to still exist. There cannot be two pasts. If you could travel back in time to prevent your birth then you would not be here to travel backwards in time for that purpose.

Usually we think of the passage of time as linear. In time travel this line closes up into a circle (Figure 5.7). Imagine a straight line of people walking one behind the other. There is a clear notion of who is behind and who is ahead of somebody else. This is like linear time: you can always say unambiguously whether an event lies in your future or your past.

Now suppose the line of people are walking around in a circle. *Locally* it appears clear that somebody is ahead of or behind you, but *globally* the ideas of 'ahead of' or 'behind' do not have any meaning when you think about the whole circle – any one person is both ahead of everyone and behind everyone else. It can no longer be said that anyone is ahead of or behind someone else. There is just a definite order.[24]

And so it is with a time-travelling history. There is no unambiguous notion of past and future. There is just one logically consistent sequence of events along a closed loop of time. It is what it is and it was what it was. You can be part of the past but you can't change it. Your experience will periodically recur if you live long enough.

Here is an example of a consistent time-travelling history. Imagine that you travel back in time and prepare to shoot yourself when you were a baby. You are determined to create a paradox of fact in the universe. You take aim at yourself when you are being held as a babe in your mother's arms. You move to pull the trigger but an old injury to your shoulder, caused by your mother dropping you when you were a baby, sends a spasm down your arm and causes your shot to miss its target. But the sound of the gunshot is enough to startle your mother, who drops the little baby on the ground, injuring his shoulder. Consistent histories make universes safe for historians.

Figure 5.7 Marchers following each other in a straight line and in a circle. In the straight line everyone is either in front of or behind everyone else. In the circle everyone is both in front of and behind everyone else.

Gödel was led to discover his rotating universe by a desire to show that the passage of time is not objective: there is no absolute standard of time. His universe is strange because it always looks the same from every point – there is constant rotation and no expansion – yet no external standard against which the rotation can be gauged (because the universe is all there is). Gödel worked out carefully how fast you would need to go in order to complete a closed trip in time, and spent much time gathering data about the distribution of galaxies on the sky because he was convinced that the universe would turn out to be spinning. Alas, our universe is not Gödel's universe. It is expanding and if it rotates at all it does so very slowly. This is easy to check because radiation coming towards us from different directions would undergo different changes to its intensity and any such differences can be no greater than one part in 100,000.[25] Just as the rotation of the Earth leads to a flattening of its shape so the rotation of the universe creates a distortion in the temperature profile of the incoming radiation on the sky, hottest along the rotation axis and coolest at right angles to it.

Gödel's discovery may not have provided a viable description of our own expanding universe but it led to new expectations about the extraordinary things that might be hidden in Einstein's equations. Universes could possess bizarre properties globally despite looking entirely innocuous locally. Gödel's universe showed that rotation could distort space in such an extreme way that time histories were closed up. Gödel showed that there were universes allowed by Einstein's equations which possessed features that had no Newtonian counterpart.

Unfortunately, Gödel never published work on cosmology again. His attentions were taken up by the hardest problems in logic and philosophy. Cosmologists spent a long time trying to find out quite how he had come to find his solution but Gödel had covered his tracks quite carefully. The strangest man.

6 The Steady Statesmen Come and Go with a Bang

Once a photograph of the Earth, taken from outside, is available, we shall, in an emotional sense, acquire an additional dimension . . . once let the sheer isolation of the earth become plain to every man whatever his nationality or creed, and a new idea as powerful as any in history will be let loose. And I think this not so distant development may well be for good, as it must increasingly have the effect of exposing nationalistic strife. It is in just such a way that the New Cosmology may come to affect the whole organization of society . . . The whole spectacle of the Earth would very likely appear to an interplanetary traveller as more magnificent than any of the other planets.

Fred Hoyle[1]

A UNIVERSE THAT ALWAYS WAS, IS AND IS TO COME

As a scientist, I simply do not believe that the Universe began with a bang.

Arthur S. Eddington[2]

In 1948, the world was recovering from the most destructive war in the history of the human race. Many scientists were returning to the work they had pursued in peacetime, while others sought to redeploy military technologies, like radar and nuclear physics, for purely peaceful scientific purposes. Sometimes, close friendships and collaborations born out of wartime projects sowed the seeds for continuing scientific work in the years that followed. Such was the case with three famous names in modern cosmology: Hermann Bondi, Fred Hoyle and Tommy Gold. Bondi and Gold were both born in Vienna and came by separate routes to Cambridge to pursue their studies. Bondi was an exceptionally talented schoolboy who had been introduced to Eddington in 1936 during the great astronomer's visit to the Austrian capital. Eddington was impressed by the sixteen-year-old and helped him apply to Trinity College, Cambridge, as a foreign student. Gold was less academically ambitious and at his father's insistence came to Cambridge in 1937 to study engineering. The outbreak of war saw both of them interned because of their nationality, briefly in Bury St Edmunds barracks, and then for fifteen months in Quebec, where they got to know each other. Released from internment in 1942 and back in England, Gold was added to an Admiralty radar research group led by Hoyle, based at Witley in Surrey. There he made several crucial contributions to the war effort, devising radar guidance for the landing craft taking part in the D-Day operation and revealing that German U-boats were using tubes to take air on board without needing to surface.

Hoyle had followed a conventional educational path in England, from Bingley Grammar School to Cambridge, where the successful undergraduate was assigned as a reluctant research student to an even more reluctant supervisor, the uncommunicative Paul Dirac. Hoyle remarked that they were perfectly suited, a student who didn't want a supervisor with a supervisor who didn't want a student. Despite early forays into the study of nuclear physics and quantum mechanics, Hoyle's interest was captured by the new subject of astrophysics. His friend Ray Lyttleton convinced him that in astrophysics there was an abundance of fascinating unsolved problems and an absence of geniuses like Paul Dirac!

Bondi, Gold and Hoyle found themselves back in Cambridge after their wartime work. Bondi and Hoyle were working on problems of astrophysics and were well known among astronomers. Gold had followed a very different research path and had become an expert on the biophysics of human hearing, using some of the insights he had gained from his work on signals with Hoyle's group. He perplexed the experts who worked in this field by adopting a very different, entirely unmathematical, approach. Although this work[3] earned Gold a position in Cambridge, it never made much impact on the medical world at the time: only in the 1970s would his predictions about the existence of feedback mechanisms within the cochlea to create acoustic resonances be shown to be correct.

Neither Bondi, Gold nor Hoyle had followed the development of cosmology until then. Bondi became interested in general relativity and quickly rediscovered the non-uniform universe that Tolman had found in the 1930s. But as they caught up with the literature and engaged in long discussions over tea and at dinner together, they came to share some dissatisfaction with the picture of an expanding evolving universe that Lemaître had impressed upon the world of astronomy. They were dismayed by the idea of a universe that began at some finite time in the past, was continuously changing, and was destined to become lifeless and barren in the future.

Following Einstein's early lead, they assumed that the universe was on the average the same everywhere in space, yet they didn't accept that it was very different in the past to what it is now, or how it will be in the future. In response to this discontent they came up with a very new cosmological scenario: a universe that was on the average the same at all times as well as in all places. Einstein's much-used assumption that the universe was the same in all places was called the Cosmological Principle by Milne. The more constraining assumption that it is also the same at all times was called the Perfect Cosmological Principle by Bondi, Gold and Hoyle.

The concept they worked with was of a situation that was steady rather than static. They knew that the universe was observed to be expanding so it could not be static. But that didn't mean that it could not be always the same. Look at one spot on the surface of a steadily flowing river and it will always look roughly the same. 'Could the universe be like this?' they asked themselves.

They wanted the universe to be on average the same at all times. There would be little local changes, stars and planets forming now and then, but all the gross properties of the universe – its rate of expansion, the rate of galaxy and star formation, the density of matter and the temperature of radiation – had to be constant on average over long periods of time. This was quite different to the standard models of the expanding universe. They had high density and temperature in the past and the rate of expansion got lower as the universe aged and headed towards a cold and lifeless future (misleadingly called

Figure 6.1 Tommy Gold (*left*), Hermann Bondi and Fred Hoyle (*right*) at a conference in Cambridge in about 1960.

the 'heat death') or a cataclysmic 'big crunch'. In the universes of Friedmann and Lemaître, there was a real difference between cosmic past and cosmic future.

In order to meet their stringent new requirement, in 1948 Bondi, Gold and Hoyle (Figure 6.1) introduced a radical new idea: if the universe is expanding then the only way in which the density of matter can stay the same, as time passes, is if new material is being continuously created so as to exactly counterbalance the dilution in density created by the expansion.

This idea sounded rather fantastic, but they argued that it was really no more fantastic than the alternative model, in which all the matter was apparently miraculously created everywhere at one moment in time. If anything, their continual creation was less fantastic and the appearance of matter might even be observed by physicists. The creation rate required to keep the density of the universe constant was actually very, very small – about one atom in every cubic metre of space every 10 billion years. This was far too small for any direct experiment to see. No physics laboratory can make a vacuum that is as empty as the average density of the universe. Yet although you couldn't hope to see the continual creation process occurring, this theory was far from being outside the reach of astronomers to test. Quite the opposite: it made lots of firm predictions that set itself up for being tested to destruction by the observational astronomers.

If the universe really displayed the same gross properties at all times in cosmic history, then when we look out in space to ever greater distances (and so see light arriving from earlier and earlier times in the universe's history), we should find things are always about the same. There can't be a time period in the past when all the galaxies started to form and before which there were no galaxies. In the 1950s, it was growing evidence against this important prediction that led to gradual disenchantment with the steady-state theory of the universe. Radio astronomers argued that galaxies which emitted radio waves were not equally prevalent at all times in the past. Later, the discovery of a new type of bright object, dubbed a quasar (short for 'quasi stellar radio source'[4]), which gave out as much energy as an entire galaxy from a region the size of our solar system, added to the problems of the theory. Quasars all seemed to be found at a particular range of distances from us, meaning that they were formed only during a particular short interval of cosmic history. This could not be explained by the steady-state theory but was entirely natural in the evolving universes of Friedmann and Lemaître, where objects like galaxies, radio sources and quasars would only begin to form when the universe had expanded enough for conditions to be suitable for that to happen.

There was also a strange coincidence which nobody noticed at the time. In the Friedmann–Lemaître universes, the expansion rate of the universe is roughly equal to the reciprocal of the age of the universe. In the steady-state universe the expansion rate has no connection with the age of the universe because the age is infinite – the universe has always existed. The expansion rate of a steady-state universe can take any value. It doesn't really matter. The fact that the reciprocal of the value found for the expansion rate of the universe is very close to the age of a typical star like the Sun is therefore a complete coincidence in the steady-state universe. Yet in the evolving universe of Friedmann and Lemaître it is completely natural. There can't be any astronomers until there are stars so it is expected that we will be observing the universe at a time when its age is close to that required for stars to settle down and burn hydrogen in a steady way (about 10 billion years)

– hence we should find the expansion rate of the universe to be a little less than the reciprocal of that number.

The version of the steady-state theory proposed by Bondi and Gold in 1948 involved very few equations.[5] It appealed to symmetry and principles of uniformity in time and space. They maintained that the universe should look roughly the same at all times from all places within it. This was an extension of the outlook, usually attributed to Copernicus, that we should not find ourselves to be in a special cosmic location. But now the location was in space *and* time not just in space.

This stipulation is extraordinarily powerful. It permits only four types of universe. The first is a static universe that contains nothing at all – no matter, no radiation, no gravity – just space and time. This is not too interesting. The next is Einstein's original static universe. It doesn't expand, so it is the same everywhere at all times. Bondi and Gold rejected this option because it conflicts with our own existence. Life needs some change to occur. Disorder and complexity must increase overall in accord with the famous second law of thermo-dynamics. The third option was another non-expanding universe: the remarkable rotating universe that Gödel had found. It rotates at a steady rate and looks the same wherever you are within it. This spin-ning universe is clearly not the one we live in, though. The last candidate was the one that met Bondi and Gold's requirements. It was the universe first found by de Sitter, with no curvature of space. Unlike the other three, it expands, but it does so at a constant rate that is the same at all times and in all places. No astronomer could determine 'when' they were living in a steady-state universe by making astronomical observations.[6] It has no beginning and no end.

Hoyle also predicted this model for his version of the steady-state universe, published shortly after Bondi and Gold's[7]. He had introduced a process of steady creation of matter which exactly counterbalanced its dilution by the expansion. This process automatically led to de Sitter's universe.

This approach had some interesting features. First, it explained why the universe was spatially uniform and expands at the same rate in all directions. If you introduce any asymmetries into a steady-state universe then their effects rapidly diminish and the expansion quickly resumes its uniform and isotropic state of expansion. The steady-state universe is stable against the effects of small disturbances.[8] If, for example, you balance a pencil on its point momentarily, then you will soon discover that it is unstable in this position because any small perturbation will make it fall over. But place a marble at the bottom of a kitchen mixing bowl and push it away from its resting place at the bottom and you will find it quickly oscillates back to the starting point. It is stable.

The steady-state universe created a lot of controversy because it seemed at first to be abandoning Einstein's theory of gravity in order to describe the universe on the basis of symmetry principles alone. Bondi, Gold and Hoyle didn't seem to worry too much about this because their theory could make a lot of simple predictions that would allow it to be tested by the astronomers. Hoyle's phenomenon of continuous creation appeared to be a piece of new physics that was not part of the general theory of relativity. But a few years later, in 1951, the British astrophysicist William McCrea showed that Hoyle's creation field could easily be added to Einstein's theory without changing its character. It was just another way of expressing the cosmological constant which Einstein had introduced long ago. The cosmological constant just gave the rate of creation of new material. It was constant always and so it gave rise to de Sitter's universe, shown in Figure 3.4.

Throughout the 1950s the steady-state theory was regarded as a serious rival to the conventional 'big bang' universes, as Hoyle had dubbed them in his 1949 radio broadcasts about the astronomical universe on the BBC Third Programme. The new frontline of observational astronomy was radio astronomy, spearheaded by the telescopes at Cambridge put in place by Martin Ryle and using new techniques that he had developed to improve the quality of astronomical observations. Although they had both worked on radar development during the war, Hoyle and Ryle didn't get along and each suspected the other was more interested in proving him wrong, with Ryle trying to disprove the steady-state theory's prediction and Hoyle undermining the accuracy and interpretation of Ryle's surveys of radio sources in space. In the end, Ryle's data convinced almost everyone that astronomical sources of natural radio waves were not equally prevalent in the universe at all times in the past as the steady-state theory required. The following poem, written by Barbara Gamow, the wife of George Gamow, captures an outsider's view of the dispute:

Ryle versus Hoyle

'Your years of toil,'
Said Ryle to Hoyle,
'Are wasted years, believe me.
The steady state
Is out of date.
Unless my eyes deceive me,

My telescope
Has dashed your hope;
Your tenets are refuted.
Let me be terse:
Our universe
Grows daily more diluted!'

Said Hoyle, 'You quote
Lemaître, I note,
And Gamow. Well, forget them!
That errant gang
And their Big Bang –
Why aid them and abet them?

You see, my friend,
It has no end
And there was no beginning,
As Bondi, Gold,
And I will hold
Until our hair is thinning!'

'Not so!' cried Ryle
With rising bile
And straining at the tether;
'Far galaxies
Are, as one sees,
More tightly packed together!'

'You make me boil!'
Exploded Hoyle,

His statement rearranging;
'New matter's born
Each night and morn.
The picture is unchanging!'

'Come off it, Hoyle!
I aim to foil
You yet' (The fun commences)
'And in a while'
Continued Ryle,
'I'll bring you to your senses!'[9]

Bondi, Gold and Hoyle were unmoved by the astronomical evidence from Ryle's radio telescope in Cambridge and fought a strong rearguard action against the radio astronomers. The whole debate came to characterise cosmology as the 'big bang versus steady state' dispute, although in reality it was predominantly a British affair and didn't influence significantly American astronomy.[10] Indeed, it revealed an interesting difference of approach between cosmologists in Britain and America. British cosmologists took matters of scientific methodology and their underlying philosophy of science seriously in their attempts to understand the status and implications of a concept like the Perfect Cosmological Principle of Bondi, Hoyle and Gold. In contrast, the more pragmatic American style eschewed such debates.[11] So pervasive was the controversy in Britain owing to Hoyle's successful radio broadcasts and popular books that even today non-astronomers ask about the steady-state theory as if it were still contending with the big-bang scenario.

During the 1980s, I spent some time with Fred Hoyle when we were billeted together while attending a conference in Bologna about the history of modern cosmology. We talked quite a lot about this early period of uncertainty about the steady-state theory and how he viewed the evidence against it twenty years on when everyone else had abandoned the theory. What seemed to worry him most was that his process of continuous creation should produce equal amounts of matter and antimatter. This democracy between matter and anti-matter[12] looked like a problem to him at the time because there was no evidence for any anti-atoms, anti-planets or anti-stars: the universe was observed overwhelmingly to contain matter rather than anti-matter. True, we could create anti-particles in laboratory experiments and anti-electrons were routinely detected in cosmic ray showers, but the universe was predominantly made of matter, at least in our vicinity.

It is important to remember that from 1948 to 1952 all models of the expanding universe were handicapped by a miscalculation of the distances to the most distant receding galaxies, and hence significantly underestimated the age of the universe. This created the paradoxical situation that for a while the expanding universe seemed to be slightly younger than the oldest stars it contained! It was in this confusing situation that the steady-state theory arose. As a result the rival big-bang theories were in a difficult situation. They didn't quite work as detailed descriptions of the astronomical universe, and so they failed to provide a better scenario as an alternative to the steady state. Einstein's reaction to this dilemma was interesting – and, in retrospect, perfectly correct. When confronted with the dilemma that the ages of the stars appeared to be greater than the age of the universe he concluded that the theory of the evolution of the stars was more likely to be wrong: 'It seems to me, however, that the theory of evolution of the stars rests on weaker foundations than the field equations.'[13]

The situation began to change in 1952 when Walter Baade, a German astronomer who had lived and worked in America since 1931, using the 200-inch telescope on Mt Palomar began to express his doubts about the true brightnesses of the Cepheid variable stars that had been used to calibrate the distances to other galaxies when deducing the Hubble–Lemaître law for the recession speed versus distance of distant objects. His suspicions had been roused by the fact that the existing scale of distances seemed to suggest that the Andromeda galaxy, our nearest companion galaxy, which looks so much like our own, was much smaller than the Milky Way. Baade smelt a rat and tracked it down by meticulous work. He showed that the Cepheid stars were intrinsically much brighter than had been thought. As a result, astronomers had been underestimating the distances to far-away stars and galaxies quite considerably. Baade showed that everything was twice as far away, and the expansion of the universe had been going on for twice as long, as had been thought – 3.6 rather than 1.8 billion years. There would be further revisions of the distance scale in astronomy, and by 1956 the likely age had risen again to 5.4 billion years. This continuing uncertainty has only been significantly over-come in recent years by the discovery of new distance indicators for very distant objects and the observational capabilities of new ground-based and space telescopes like Hubble and Spitzer. Today, the best estimate of the expansion age of the universe is 13.7 billion years with an uncertainty of just 0.1 billion years.

A TABLE-TOP UNIVERSE

In the early days we usually had to work on a very small budget. We certainly were very poor. In the 1941 paper, for instance, I had to do all the work in person, including the rebuilding of the labo-ratory room and making all the electrical installations. It was time-consuming, but gave some additional spices and satisfaction when you found that you could handle everything yourself. The scientist today is certainly a rather spoiled person, especially in the rich countries.

　　Erik Holmberg[14]

Today, computers are so familiar and easy to use that it is hard to imagine a world where it was otherwise. Astronomers use large computer simulations to follow the complicated chain of physical processes that have led to the formation of stars and galaxies, and the special patterns they weave on the sky. In 1941, Erik Holmberg, a young Swedish astronomer at Lund Observatory performed the first simula-tion of a universe. He didn't use a computer in the modern sense (they didn't exist); instead, he created an analogue device that mimicked the behaviour of gravity and then watched and measured what it did when many gravitational forces from different stars were allowed to act together.

　　One of the striking features of living in a world of three dimen-sions is that many basic forces and effects in Nature fall off in strength in inverse proportion to the square of the distance you are from their source.[15] This is true for the strength of gravity, magnetism, electro-static forces and light intensity. Holmberg wanted to understand how two galaxies of orbiting stars might interact if they passed close to each other. How would the pull of the other's gravity distort the orbits of the stars and the shapes of the galaxies? In an expanding universe, galaxies would have been closer in the distant past, and many more close encounters would have taken place than today.[16]

Holmberg decided to exploit the fact that light intensity falls off with distance in the same way as gravitational force. By rigging up an array of light bulbs, he could mimic the distribution of gravitational forces that would be created by a system of masses. He set up two models of flat-disc galaxies (like the Milky Way) by laying out two sets of concentric circles of bulbs, thirty-seven in each, on a flat table with a dark base. The brightness of each bulb was proportional to the mass of the star it represented and their brightnesses were adjusted according to their distance from the centre in proportion to that expected in real populations of stars. Every bulb had a sensor on each of its four sides, so that the total light falling on each of its sides from the other bulbs at different distances could be measured by a photocell and a galvanometer. These bulbs were unusual and specially made for Holmberg at a local factory. Each 'galaxy' of light bulbs was given a motion consisting of an overall rotation plus a straight-line motion towards the other galaxy to start with. Holmberg would then measure the light intensities falling on each bulb and work out the net effect (in magnitude and direction). This would tell him how the stars would move. All the seventy-four light bulbs were then moved to new positions and the new brightness variations measured to give the new 'forces' that dictated the next move of the bulbs. In this ingenious fashion Holmberg was able to create an analogue model of how two systems of stars would interact with each other by the force of gravity, and follow their evolution step by step – a problem far too complicated to solve exactly using pencil and paper and Newton's equations alone.[17]

Figures 6.2 and 6.3 are from Holmberg's paper.[18] He detected the tendency for the two galaxies to become most deformed in shape after their moment of closest approach, when their gravitational attraction was greatest, and they often produced armlike features reminiscent of the spiral arms seen in many disc galaxies like the Milky Way. When the two galaxies both rotated in the clockwise sense the arms arched outwards, pointing in the direction of rotation, but pointed in opposing directions when the two galaxies were set rotating in different senses.

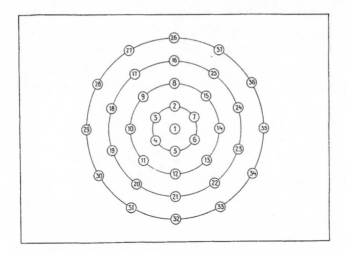

Figure 6.2 Erik Holmberg simulated galaxies using electric lamps. Here, he shows the lamps they are arranged in a concentric group 80 cm in diameter, containing thirty-seven lamps. He made two of these groups; each represents a galaxy of stars. Voltages across the lamps can be adjusted to give any desired distribution of 'mass' (i.e. light) within each group. The diameters of the light bulbs were 8 mm, far smaller than the distances of 10–20 cm between the lamps.

He also concluded that medium-sized and small galaxies would often be captured by the very strong gravity pull of a big galaxy, then merge, and produce a single big galaxy of an elliptical rather than a spiral type.

Holmberg's imaginative work introduced a new tool into the study of the universe. It would not re-emerge until the 1970s, but then it grew in scope and power along with the advancing computer revolution. That new tool was the ability to build computer models of the universe, or simplified versions of it. Programmed with the simple law of gravity, which could calculate the gravitational forces between any number of masses introduced into an expanding model, the computer could follow their future courses, look at how they clustered, and determine when galaxies first formed and perhaps collided with their nearest neighbours. Eventually, we hope to be able to follow all the messy details of star formation within those galaxies.

Today, cosmologists simulate different types of universe on some of the world's biggest computers, predicting what we should see

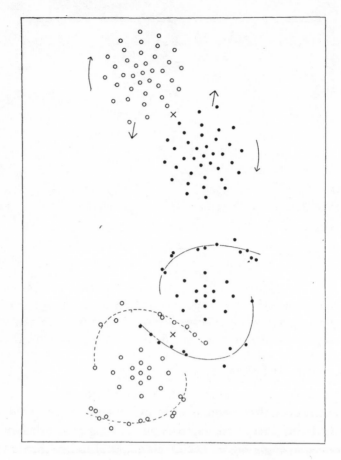

Figure 6.3 Two of Erik Holmberg's electric-light 'galaxies', both rotating in a clockwise direction, approach to within a distance equal to each of their diameters. Their gravitational interaction leads to spiral arms being created, which point in the direction of the galaxies' rotation. The short arrows show the direction of the galaxies' motion in space near the point of closest approach. The curved arrows show the direction of rotation.

through our telescopes, or infer from the statistical analysis of our data. Whereas once there were two types of astronomer, the observer and the theorist, now there is a third. The computational astronomer has different skills, which can produce intricate computer codes and impressive graphics or video to extract and display the consequences of different possible universes.[19] Theorists are as interested in making predictions for computer modellers to test as they are for observers to search them out with their telescopes.

THE ELECTRIC UNIVERSE

... what with having perforce to change a light bulb here and
tune in a transistor radio there, I have picked up a pretty sound
working knowledge of electrical matters. It is not comprehensive,
God knows – I still can't fully understand why you can't boil an
egg on an electric guitar.
    Keith Waterhouse[20]

During the same period that he was developing the steady-state theory
of the universe in 1948, Hermann Bondi proposed another, quite
different type of universe. It might have the steady-state form, but
that was neither necessary nor conditional upon the source of the
expansion that he imagined. The idea was developed with his
Cambridge colleague Ray Lyttleton and exploited the fact that the
electrostatic force of repulsion between two charged particles like
protons is vastly greater (10^{39} times greater, in fact) than the gravitational force of attraction between them.

Usually, every hydrogen atom is assumed to be electrically neutral.
It contains one proton and one electron, whose charges are assumed
to be exactly equal in magnitude but opposite in sign: the proton
charge is $+e$ and the electron charge is $-e$, so their sum is zero and
the atom is electrically neutral. But suppose, Bondi and Lyttleton
suggested, that there was a tiny undiscovered difference between the
magnitude of the proton and electron charges. Hydrogen atoms would
have a net overall electric charge of the same sign and would repel
each other if the repulsive force was stronger than the force of gravitational attraction between them (Figure 6.4).

The electric repulsion is so much stronger than gravity, they argued,
that if the magnitude of the proton's charge was greater than that of
an electron's charge by merely 10^{-18} of the basic charge, e, on the
electron then the resulting repulsion between atoms would be large
enough to explain the expansion of the universe!

When this novel idea[21] was proposed, the widely known experimental constraints on this charge imbalance were 100 times weaker
(less than 10^{-16}) than Bondi and Lyttleton proposed. Then, gradually,
it emerged that there were more sensitive experimental limits of

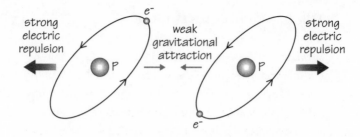

Figure 6.4 Two hydrogen atoms exhibit very weak gravitational attraction. But if the charge of the proton (p) is not equal to the charge of its orbiting electron each atom will have a non-zero net charge. These excess charges will be the same for the two atoms and will create a strong electric repulsive force between them.

$10^{-20}e$ on the allowed charge imbalance. At first, Bondi and Lyttleton resisted these results because they maintained that the charge imbalance might have subtle effects on the atomic structure and electromagnetic fields involved in the experiments themselves that had not been accounted for when the results were interpreted. But over time these fears were allayed and the theory could no longer be defended in the face of the experimental evidence.[22] Finally, studies of the charge of hydrogen molecules, composed of two hydrogen atoms, by John King in America in 1960, found that the maximum-allowed charge imbalance was at least forty times smaller than was needed to explain the expansion of the universe.[23]

This theory was never of major interest to cosmologists but it was an early example of how the injection of new sub-atomic physics into the picture of the expanding universe could change the story told by Einstein's theory of gravity.

On the other side of the Atlantic a different model of the universe was being pursued by a small group of cosmologists who were more interested in the physical processes that happen in the early stages of an expanding universe than in solutions to Einstein's equations.

HOT UNIVERSES

'Some like it hot.'
   Billy Wilder and Iţec Diamond[24]

The steady-state picture of the universe attracted so much publicity in Britain that its rival almost slipped from public view. Yet, after 1948, there was increasing interest in the big-bang model and what might have happened in a universe that was once hotter and denser than it is today. Whereas astronomers had focused upon distinguishing the different types of expanding universe that Einstein's theory predicted, physicists started to become interested in the unusual environment of the early universe. Lemaître had pioneered thinking about the physical state of the embryonic universe but it was the Russian émigré George ('Geo') Gamow (1904–68), and his students, who pushed forward a pathway that was to prove the most useful and trustworthy guide to the past.

Gamow was a colourful character who lived an exciting life in exciting times. He grew up in Odessa against the turbulent backdrop of the Russian Revolution, studied in St Petersburg, where he attended lectures on general relativity by Alexander Friedmann, and made important contributions to nuclear physics while still in his early twenties. He was an extrovert and unconventional individual who seemed to have a knack of encountering people of every conceivable hue, from Molotov, Bukharin and Trotsky to Einstein, Bohr and Francis Crick, usually in very memorable circumstances.[25]

Gamow's departure from his native Odessa for the United States was suitably unconventional. Appalled by the growing interventions of the communist state into intellectual life, in 1932 Gamow and his wife Rho decided that they would escape from the country by crossing the least defended border. They picked the southern tip of the Crimean peninsula and hoped that smugglers would help them across the Black Sea to Turkey, just 170 miles away. As part of their master plan they acquired a collapsible rubber-hulled kayak of the sort used on expeditions, which they planned to carry on their backs along with oars in two rucksacks. Despite the food shortages in the Soviet Union at the time, they managed to build up a week's supply for the trip and added some improvised flotation devices (football bladders), a water pump and a compass. They aimed to arrive without ID cards in Turkey, claiming to be Danes on the basis of an old Danish driving licence Gamow owned, and hoped to end up at the Danish embassy in Istanbul. In early summer 1932, they booked holiday rooms in the Scientific Academy's vacation residences on the Crimean coast, taking the boat out and learning to paddle and manoeuvre it.

Originally, they had planned to wait until the full moon before making good their escape but they now decided to go while the sea looked calm. For the first two days the sea journey went smoothly but then the wind and waves grew stronger, swamping the boat so that Rho had to work the pump continually to stop them from sinking. When the wind finally dropped they found themselves within reach of the shore and in sight of fishermen. Sadly, they were not Turkish: the Gamows were only seventy miles from where they had started out and were taken back to their starting point by the helpful fishermen. Fortunately, no one doubted their story that they were merely kayakers caught out by the weather and swept far from their home beach. The real truth would have been too crazy to be believed.

The Gamows didn't give up their ambition to leave. When two years later, helped by an invitation from Niels Bohr, they attended a physics congress in Belgium, they didn't return to the Soviet Union. Geo was offered a position in Ann Arbor, Michigan, and would spend the rest of his career in the United States.[26]

Using his deep knowledge of nuclear physics Gamow imagined the whole universe behaved like a huge nuclear physics experiment during its first few minutes of its expansion. After some false starts, the key ideas began to be assembled in the right order by two of Gamow's young research students, Ralph Alpher (1921–2007) and Robert Herman (1914–97), in 1948.

In the summer of 1948, Gamow had shown how the nuclear combination of a proton with a neutron could produce deuterium when the universe was about 100 seconds old[27]. Six months later, Alpher and Herman, in a letter to the journal *Nature*, developed Gamow's idea further, calculating how the temperature and densities changed during the history of the universe so that they could link their values today with those at any given time in the early history of the universe.[28]

Alpher and Herman showed that the ratio of the matter density of the universe to the cube of the temperature of any heat radiation present from its hot beginning is a constant if the expansion is uniform and isotropic. This meant that they could determine what the value of this ratio needed to be when the universe was two minutes old, and its temperature a billion degrees, in order to avoid making more helium by nuclear reactions than we see today. Then, knowing the required value of the ratio and the observed value of the matter density today, they could deduce what the current radiation temperature should be. They estimated it to be about 5 degrees Kelvin. This was one of the most momentous predictions ever made in science. It offered astronomers a way to test if the universe had once been as hot and dense as the Big Bang theory predicted. It showed that there should be radiation 'fall-out' in the universe today if it emerged from a high-temperature past.

Alas, no one seemed to take any notice of these papers. The steady-state controversy diverted people's attention away from the alternatives in 1948 and no one had heard of Alpher and Herman in Europe. Things weren't much better in the United States. In the mid-1960s a leading physicist, Robert Dicke at Princeton, would begin a major programme to search for this radiation without having read the papers in the world's leading journals by Gamow, Alpher and Herman that made those early predictions.

Two top-flight radio engineers, Arno Penzias and Robert Wilson, working on calibrating a receiver to track the Echo communications satellite for Bell Labs at Holmdel, New Jersey, would eventually detect this radiation from the big bang in 1965 without knowing what it was – or that cosmologists like Alpher, Herman and Gamow had predicted that such radiation should exist. Their detection of radio noise at a wavelength of 7.35 cm, equivalent to a heat radiation temperature of 3.5 ± 1.0 degrees Kelvin was soon made known to Dicke and his team, who had been trying to estimate this radiation temperature. One of his former students, James Peebles, submitted a paper estimating the expected temperature to be about 10 degrees. This paper was eventually rejected by the journal's referee (who was Alpher) on the ground that it was not original and the calculation had been done before by Alpher and Herman. However, Dicke's team still didn't seem to take this information on board and pressed ahead by publishing an interpretation paper alongside Penzias and Wilson's low-key announcement of the radiation detection in the *Astrophysical Journal*. Much has been

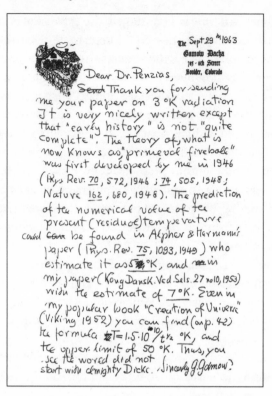

Figure 6.5 Letter from George Gamow to Arno Penzias in 1965 (dated 1963 by Gamow!) pointing out the earlier prediction of the background radiation after Penzias and Wilson announced their discovery alongside a theoretical interpretation by Robert Dicke and his collaborators.[29]

written by historians of science about this failure of collective memory that set the subject back many years. In 1975, in his autobiographical notes, Dicke wrote that:

There is one unfortunate and embarrassing aspect of our work on the fireball radiation. We failed to make an adequate literature search and missed the more important papers of Gamow, Alpher and Herman. I must take the major blame for this, for the others in our group were too young to know these old papers. In ancient times I had heard Gamow talk at Princeton but I had remembered his model universe as cold and initially filled only with neutrons.[30]

Alpher and Herman remained unhappy about their treatment for the rest of their lives and played only a minor part in the future of cosmology after they took jobs in industry.[31] Alpher left Johns Hopkins University to work for General Electric in 1955 and died in 2007. Herman went to lead the physics section of the General Motors Research Laboratory in 1956, before becoming a professor at the University of Texas, where he did award-winning work on the mathematical theory of traffic flow and became an accomplished sculptor of wood minia-tures, which are on display in several US galleries. He died in 1997.

Although neglected for a long time, Alpher and Herman eventually received many prizes and honours for their work on the background radiation and the big-bang. Alpher was awarded the National Medal of Science in 2005, two years before his death. The citation read:

For his unprecedented work in the areas of nucleosynthesis, for the predic-tion that universe expansion leaves behind background radiation, and for providing the model for the Big Bang theory.

The detection of radiation by Penzias and Wilson was a turning point in our understanding of the universe and our confidence in the ability of Einstein's equations to predict how universes behave. The simplest expanding universes of Friedmann and Lemaître tell us what the temperature will be in the universe at every moment of time. Armed with this simple fact, physicists could map out the sequence of events that would unfold as the universe expanded and aged from being a few seconds old until the present. They couldn't determine everything that happened – far from it – but the broad picture of how the temperature and density decreased, when nuclear reactions occurred, and when the radiation cooled enough for atoms and molecules to form, was soon established.

The discovery of the relic heat radiation also sounded the death-knell for the steady-state theory. Arguments about whether the abundance of radio galaxies was really the same at all epochs in the universe's history were suddenly trumped by the discovery of fossil heat radiation that was entirely natural (and predicted) in the big-bang theory but unexplained in the steady-state theory. A strong rearguard action, led by Hoyle, was waged in defence of the steady-state theory by its supporters, who tried to show that the new background radiation could

Figure 6.6 Robert Herman (*left*), George Gamow and Ralph Alpher (*right*). Gamow is emerging from a bottle of primordial 'Ylem', the stuff of which the universe was made – although it looks distinctly more alcoholic than that.

have been produced more recently in our galaxy. These valiant attempts all failed: the background radiation had far too many photons in it compared to the density of matter in the universe to have been produced by any known sources of radiation. By 1967, it was known that the radiation had the same intensity around the sky to an accuracy of better than one part in a thousand, a smoothness that couldn't be achieved by adding together different local sources of radiation without their number becoming so large that we would see them.

Curiously, Hoyle also played a key part in establishing the hot big-bang theory. He used his expertise in nuclear physics to go much further than Gamow, Alpher and Herman in predicting the abundances of the lightest elements that should emerge from the first few minutes of the big bang. First, in a pioneering paper[32] with Roger Tayler at

Cambridge in 1964, and then in work at Cal Tech with Willie Fowler and his student Robert Wagoner in 1967, he predicted the whole spectrum of light elements (notably, deuterium and the isotopes of helium and lithium) that would be produced in different big-bang universes.

These impressive predictions were made possible by a small but crucial observation in 1953 by the Japanese astrophysicist Chushiro Hayashi (1920–2010), who is most famous for his work on the evolution of stars. All the physicists, like Gamow, who had tried to predict the abundances of the elements that would emerge from the big bang had one major problem. What are the relative numbers of protons and neutrons that the universe starts out with? All nuclei are made of combinations of these two particles (for example, helium-4 is made of two protons and two neutrons), so surely the final abundances must depend on what the starting balance is? If you start with all protons you may end up with 100 per cent hydrogen, but otherwise how can you predict what happens when you don't know how the abundances start out?

Hayashi pointed out that the weak radioactive force of Nature solved the problem. When the universe is younger than one second old and hotter than 10 billion degrees this force mediates weak radioactive interactions between protons and neutrons that keep their numbers equal. A complete equilibrium is maintained in which the number of neutrons per proton is determined *only* by the temperature. So you don't need to know anything about the 'beginning' of the universe or other such unknowable things in order to calculate the relative numbers of neutrons and protons. In fact, the neutron is very slightly heavier than the proton and as the temperature falls to around 10 billion degrees, after one second, the little bit of extra energy you need to make a neutron rather than a proton means that the number of protons starts to exceed slightly the number of neutrons. However, the imbalance doesn't get a chance to grow very much because the crucial weak interactions between the protons and neutrons become too slow for the reactions to keep up with the expansion of the universe and they stop. The ratio of neutrons to protons in the universe is then fixed at its value when this happens, and is about 1 to 6.

Nuclear reactions set in suddenly after about 100 seconds of expansion when the temperature has fallen to a billion degrees. By then, a

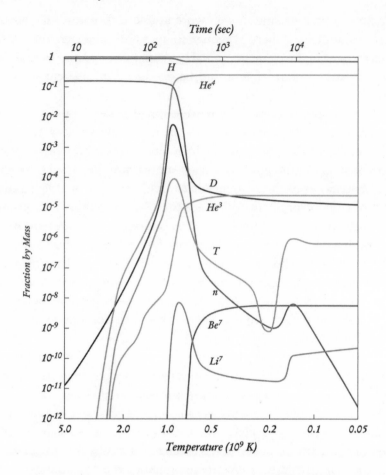

Figure 6.7 The build-up, with increasing time and falling temperature of the fraction by mass of elements in the universe. The curves show the abundances of the lightest nuclei (hydrogen (H), helium-4, deuterium (D), helium-3, tritium (T), beryllium-7 and lithium-7) produced in the first three minutes of the big-bang universe. The curve labelled *n* tracks the abundance of free decaying neutrons that were not bound into nuclei.

little bit of further neutron radioactive decay has occurred, which decreases the ratio of neutrons to protons slightly further, to about 1 to 7. Almost all the surviving neutrons quickly end up in helium-4 nuclei, leaving just a few traces of deuterium, helium-3 and lithium. The universe ends up with about 23 per cent of its nuclear matter in the form of helium-4, about 77 per cent in the form of hydrogen, with tiny traces of the isotopes of deuterium (10^{-3} per cent), helium-3

(10^{-3} per cent), and lithium-7 (10^{-8} per cent) left over as well. These are the abundances of these elements that we find wherever we look in our galaxy and beyond today, so this application of simple nuclear physics to the first few minutes of the universe's history is beautifully confirmed by astronomical observations (Figure 6.7).

Since 1975, the reconstruction of the distant past of the universe, using our growing knowledge of high-energy physics and elementary particles, has become a major new research area, known as 'particle cosmology'. It was destined to revolutionise our understanding of the early stages of the universe.

7 Universes, Warts and All

'I am the cat who walks by himself, and all places are alike to me.'

Rudyard Kipling, *Just So Stories*

TURBULENT UNIVERSES

'When I die and go to Heaven there are two matters on which I hope for enlightenment. One is quantum electrodynamics and the other is the turbulent motion of fluids. And about the former I am really rather optimistic.'

Horace Lamb[1]

With just a few isolated exceptions, the thrust of the early study of universes emerging from Einstein's theory had focused on those that were smooth and uniform, expanding at the same rate in all directions. There were two very good reasons for such a bias. On the one hand, it accorded rather well with what was observed through astronomical telescopes. There was no evidence for expansion rate differences between one direction and another, and the clustering of galaxies didn't seem to be very different from being random all over the sky. In the early days of measuring redshifts of distant galaxies this was a time-consuming process. Most of a night at the telescope would be spent measuring a few redshifts. It was not until the appearance of photometric detectors based upon CCD technology used in video cameras in the 1970s (for which the 2009 Nobel Prize in physics was awarded) that many redshifts could be measured very quickly. Large surveys of galaxy positions on the sky could now measure their redshifts (and hence deduce the distances to the galaxies), so as to create a three-dimensional picture of the galaxy distribution. The resulting three-dimensional maps were a big surprise and revealed that the projected positions on the sky were rather misleading indicators of the richness and subtlety of how galaxies were clustered into a cosmic 'web' of lines and sheets.[2]

We have seen how the introduction of 'principles' like the Cosmological Principle or the Perfect Cosmological Principle were used to give these simplifying assumptions some type of philosophical rationale, if it were needed. Yet there are also good reasons for trying to go further along the road of greater realism in descriptions of the universe. After all, the real universe does contain galaxies and other irregularities in the distribution of matter. Where do these irregularities come from? Why do they have the shapes and sizes that they do? Although deviations from perfect isotropy in the expansion seemed to be small, they cannot be exactly equal to zero if we reject the idea that we are located at a special place in the universe. If space is non-uniform it cannot be isotropic about *every* point.

Lemaître had been the first to think seriously about the origin of galaxies, and Lifshitz's paper about the stability of the expanding universe when there are small deviations from perfect uniformity gave an important indication as to why the universe contained so many 'lumps' of material in the form of stars and galaxies. If any regions contained slightly more material than the average then they would attract yet more material to them at the expense of the sparser regions. This process of gravitational instability, first understood by Newton and calculated in detail by Sir James Jeans in 1902, acts very quickly in a region of space that is not expanding or contracting. But, as Lifshitz discovered, if the region is expanding then the rate at which the irregularity grows is slowed. The agglomeration of material has to beat the overall trend of the expansion which wants to drag it apart, so this slower rate of irregularity growth is not surprising.

Cosmologists wondered if this simple process might ultimately explain why galaxies existed in the universe. Perhaps some very small 'random' variations from perfect smoothness existed at the 'beginning', or were produced by some specific process soon afterwards, and they just grew bigger and bigger until they separated off from the overall expansion and settled down to form the structures we call 'galaxies'

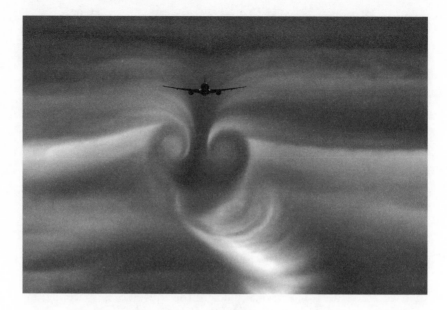

Figure 7.1 The turbulent airflow behind a jet aircraft.

today. That still leaves a lot to be explained, though. Why do the galaxies mostly contain about 100 billion to 1000 billion stars? Why do they have such a particular range of elliptical and spiral shapes? And, most striking of all, why are so many of them spinning?

During the period from 1944 to 1951, the German physicist Carl von Weizsäcker started to think about the possibility that many structures in the universe might be the residues of a rather more turbulent past history. This is a tempting thought. Turbulence is very familiar and ubiquitous. Turn on the bath tap and you witness a rushing turbulent flow of water; watch the waves lashing the sea shore, or the leaves swirling in the air behind an accelerating car, and you see it again (Figure 7.1). Yet, surprisingly, this familiar experience is one that is a virtually intractable problem for mathematicians who study the flow of liquids. When the flow is fast and swirling it is very difficult to arrive at a clear understanding of what is happening. If the flow is slow and the turbulence rather gentle then there is hope of progress but otherwise our computers and calculational abilities are easily defeated by the complexity. One of the million-dollar Clay Millennium Prizes offered for the solution of the seven greatest unsolved problems in mathematics is for the solution to this problem.[3]

Von Weizsäcker was interested first in trying to explain the origin of the solar system[4] and the motions of stars in galaxies.[5] Then, intrigued by the pictures of spiral galaxies which suggested that their whirlpool-like appearance was the residue of a previous turbulent era in the history of the universe, he proposed that turbulence was the key to understanding the existence of galaxies.[6] In this crusade he had been joined by his fellow countryman Werner Heisenberg, one of the architects of quantum mechanics, who developed a mathematical theory of turbulence and remained fascinated by the phenomenon throughout his career.[7]

George Gamow was also seduced by this picture of a once turbulent universe and put forward a similar theory in 1952.[8] But while it is easy to imagine that galaxies might emerge from a turbulent mess, it is not so easy to turn this into an exact theory. Turbulence wasn't understood in the laboratory, let alone in the expanding universe, although in the latter environment things did at least happen more slowly and vortices of material spun around in space are free of the confining boundaries, nozzles and plug-holes that make bathroom turbulence so complicated.

Lifshitz had shown that weak vortices would decay in the later stages of the universe but their velocities would stay constant during the hot-radiation-dominated phase of the universe. The reason is the same as that which explains why a pirouetting ice skater spins faster as she draws her arms in towards her body – and would spin slower if she stretched her arms outwards again. A large spinning eddy in an expanding universe is having its 'arms' stretched and as it increases in size so the rotation speed of the material falls.[9]

Turbulence has a particular simple feature: if it is created by the stirring up of large vortices in a liquid, then its energy cascades down to smaller and smaller vortices until eventually they are small enough to be dissolved by the friction of the fluid. Put a drop of ink in a glass of water and stir it very gently and you will see this cascade happening. In 1941, the great Russian mathematician Nikolai Kolmogorov had suggested that in between the largest dimensions where the turbulence first got stirred up and the smallest dimensions where it gets damped out by the viscosity of the fluid, it should always tend to transfer energy from larger to smaller vortices at a constant rate.[10] As a result, the speed of rotation of the spinning eddies in the turbulence should be proportional to the cube root of the size of their diameter[11] over a wide range of sizes, irrespective of how the turbulence originated and how it gets dissipated away on very small scales by friction. This is an appealing feature because this characteristic variation of the velocity with the size of the spinning eddies might be linked to the rotation speeds of galaxies of different sizes. But the history of a turbulent universe is a lot more complicated.

During the first 300,000 years of the universe's expansion the speed of sound is very high and the vortical motions would have been subsonic. But after that time the electrons and protons bind together to make atoms, the radiation stops interacting with the electrons and the speed of sound drops dramatically. Suddenly the vortices find themselves spinning supersonically. Cataclysmic events ensue. Huge shock waves form and matter is piled-up in large irregularities. Out of this shock-induced pile-up of spinning material galaxies were supposed to form. When everything settled down we hope to see large rotating galaxies.

This scenario offered the possibility that we might be able to explain the existence and the rotation of galaxies as a consequence of the special Kolmogorov pattern of turbulence so long as the origin of the cosmic turbulence could be explained and the supersonic pile-up of matter understood.

Gamow's support for the turbulent universe didn't attract many followers and the theory went into abeyance until the mid-1960s, when strong research groups led by Hidekazu Nariai in Hiroshima and by Leonid Ozernoy in Moscow both started to develop the theory separately. Much energy was expended on this quest from 1964 to 1978 and it remained a viable theory for the origin of galaxies until the late 1970s. It was a tricky picture of the universe to study because it was too difficult to find a solution to Einstein's equations which described a real turbulent universe. All you could do was trace the fate of vortices formed near the beginning of the universe, and check that they survived and created irregularities in density and pressure from which islands of spinning material could form and somehow end up looking like galaxies we see in the sky.

Two big problems emerged which eventually killed the whole idea. At first, it looked as though the turbulent vortices need not affect the overall expansion of the universe very much. Alas, this turned out to be untrue. Although the vortices had the same speed as you followed their history backwards in time, they had an ever-growing gravitational effect upon the expansion of the universe. Eventually, there had to be a past time before which the entire universe was a chaotic turbulent mess. In order for the vortices to spin fast enough to produce the galaxies we see today, this state of total turbulent chaos would have occurred far too recently to accord with our observations of the background radiation. It would also destroy the successful synthesis of helium, which produced the observed abundances, when the universe was about three minutes old.[12]

The second big problem for these turbulent scenarios was a radical change in what we thought about rotating galaxies. Until 1974, astronomers believed that all elliptical galaxies rotated and the faster they rotated the flatter their shape. James Binney, a fellow research student in Oxford, then showed that the flattening of elliptical galaxies didn't need rotation to explain it.[13] Their stars were most likely moving in random orbits and the overall shape was a consequence of the formation process. Gradually, observations started to support this perspective: rotation speeds of many elliptical galaxies turned out to be far too low to explain their flat shapes and some ellipticals even turned out to rotate about the wrong axis. Put together, these two developments led to the rapid demise of the turbulent theory of galaxy formation. Unfortunately, turbulence just has too drastic an effect upon the expansion of the universe and caused it to be horribly irregular and distorted in ways that can't be reconciled with the smooth and almost perfectly spherical expansion we see today.

DISTORTED UNIVERSES: FROM ONE TO NINE

The purpose of any Cosmogonic Theory is to seek out ideally simple conditions which could have initiated the world and from which, by the play of recognized forces, that world, in all its complexity, may have resulted.
  Georges Lemaître[14]

The study of universes up until 1950 had been dominated by the different varieties of universe that are the same everywhere and expand at the same rate in every direction. These are called 'homogeneous' (the same in every place) and 'isotropic' (the same in every direction) universes. They are the most symmetrical and so they are the easiest solutions to Einstein's equations to find and the simplest to understand. Occasionally, a different sort of universe would be found, like those discovered by Kasner, Einstein and Rosen or Straus, Tolman and Gödel, or the behaviour of very small distortions from uniformity and isotropy were investigated, as was first done by Georges Lemaître and Evgeny Lifshitz. But these deviant universes had been discovered in a haphazard fashion. There was no clear understanding of how many of them there might be and into what categories they might fall. This changed in 1951, when Abraham Taub (1911–99) completed a far-reaching systematic study of all the universes that are the same everywhere in space but can expand at different rates in different directions.[15] No matter where you are located in such a universe you would observe it to have the same history. Like the cat who walked by himself, all places will look alike to astronomers in these universes, wherever they are. Yet if you look in different directions you will see differences. The simplest of these was the universe first found by Edward Kasner in the 1920s, but how many other types of universe could there be that look different from one direction to another but are the same everywhere?

When I first became a post-doctoral researcher at the University of California at Berkeley in 1977, Abraham (or 'Abe' as he was known) Taub was the head of the general relativity research group in the Mathematics Department. I was in the Astronomy Department but I would go to the Mathematics Department for the seminars and informal lunches organised by Taub's research group. Taub was rather intimidating for younger scientists because he had an extraordinarily wide range of knowledge and experience, was searchingly critical and had no small talk: you didn't talk to him about half-baked ideas. He had set up the Computer Laboratory at Berkeley, worked with von Neumann at Princeton, and interacted with Einstein and Gödel; he was a world authority on shock waves and hydrodynamics, as well as an expert on the general theory of relativity and cosmology who had completed his doctorate under the supervision of Robertson at Princeton in 1935. He also ran his seminar in the old inquisitorial style, never letting any statement

Figure 7.2 Luigi Bianchi (1856–1928).

by the speaker pass if he had some doubt about it; in this he treated visiting Nobel Prize-winners the same as graduate students. At first, he seemed rather crusty, but gradually you realised that he liked to help: he treated all the young researchers as if he were the father of a large and potentially unruly family who needed to be kept on the right tracks.

In 1951, Taub had realised that back in 1898 the great Italian geometer Luigi Bianchi (1856–1928) had solved the problem of finding all the possible spaces that would be seen to be the same by observers located anywhere within them.[16] Directional differences have to look the same from every vantage point. Bianchi (Figure 7.2), who spent his entire career as a professor at several elite mathematical institutes in Pisa, showed there were just nine possibilities, which he labelled by the Roman numerals I–IX.

Taub could use Bianchi's classification to find all the types of space

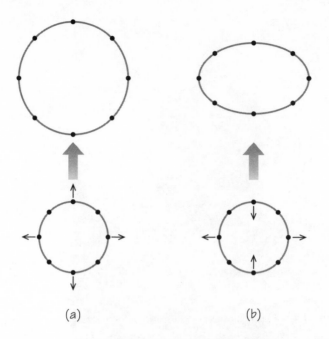

Figure 7.3 The action of shear distorts spheres into ellipsoids: (a) expansion without shear; (b) expansion with shear distortion.

that could be accommodated by expanding universes that were the same everywhere.[17] The Roman staircase of possibilities from space I to space IX that Bianchi laid out took us from the simplest flat geometries of Euclid, like Kasner's universe, to the isotropic curved spaces of Riemann and Lobachevskii, and then on to new types of space which were curved in different ways in different directions. Some could even change their curvature as a universe expanded, behaving for most of the time like a negatively curved 'open' universe but occasionally flipping to display positive curvature. Some, like Gödel's universe, could just rotate, but others could rotate and expand in a distorted way.

Taub was not able to solve Einstein's equations for all these possible universes, although he could for some of them and he wrote down Kasner's universe in the simple form that we have already seen. The most complicated of these universes are described by equations that remain unsolved to this day but the overall behaviour of almost all the possibilities has gradually become qualitatively clear over the past

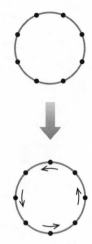

Figure 7.4 The action of rotation twists spheres.

fifty years. The 'Bianchi universes', as they have become known, started to be studied intensively in the 1960s and remained of central interest until the late 1980s, because of the need to understand why the background radiation was so isotropic around the sky.

Taub's anisotropically expanding universes can possess four key features which make them more complicated and exotic than the simple universes of Friedmann, Lemaître and de Sitter:

1. Shear distortion
2. Rotation
3. Velocities relative to the spherical Hubble expansion
4. Anisotropic curvature

The first of these properties corresponds to taking a spherical ball and squeezing it top and bottom so that it changes into an egg shape. Whereas an expanding isotropic universe is like a sequence of spheres of increasing size, a sheared universe is like a sequence of ellipsoids with increasing volume (Figure 7.3). Universes that rotate have to possess a shear distortion in their expansion as well (Figure 7.4). This combination of properties is fairly easy to visualise. The third ingredient is more subtle. Imagine an expanding ball. One way to trace the expansion is to draw straight lines radiating outwards from

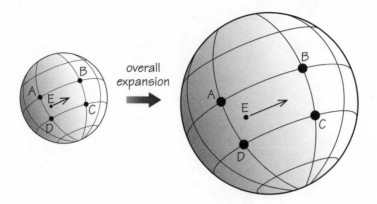

Figure 7.5 An isotropically expanding universe but with a galaxy at E moving relative to the overall expansion.

the centre. By measuring the length of those lines, AB or BC say, we could use their length as a measure of the elapsed time since the expansion started (Figure 7.5).

We have tended to assume that *we* (and all other observers on their stars in their galaxies) are also moving along those imaginary radial lines. Yet there is no reason why they should be following the overall expansion – what cosmologists call the 'Hubble flow'. Our cluster of galaxies might be moving outwards in a different direction, or there might be a variety of other random motions going on. If those other motions are relatively slow, like ours around the solar system and the solar system around our Milky Way galaxy, or involve small amounts of matter, then they will have no discernible effect on the course of the universe's overall expansion. But if they are large, with huge amounts of material moving rapidly relative to the average Hubble flow, then this can start to affect the expansion of the universe and it will develop a shear distortion.

A simple analogy makes the idea clearer. Suppose our expanding universe has just two dimensions of space and is like the surface of a balloon. If sleeping ants are put on the balloon surface and it is inflated then they will move apart in a steady fashion. They are taking part in the expansion of their universe – the surface of the balloon. But if the ants wake up and start walking around while the balloon inflates then they have an additional motion relative to each

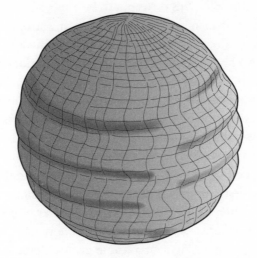

Figure 7.6 A space with anisotropic curvature can behave like an open universe in some directions and a closed universe in others. It is possible for these directional curvatures to change from positive to negative, and vice versa, with the passage of time.

other that is not simply due to the overall expansion of their space. If too many of them walk to one place they will cause a depression in the surface of the rubber and change how it expands in that direction.

The last of our four features is a brand new possibility that Einstein's theory offers. It has no counterpart in the simple Newtonian picture of the universe as a very large expanding ball. The three effects we have just described can all arise if the expansion starts at different rates in different directions, as if the elasticity of our inflating balloon was different in different directions.

We have already seen that the essence of Einstein's theory of gravity is that the motion of mass and energy change the geometry of space and endow it with curvature. In the simple isotropic universes found by Friedmann and Lemaître this curvature, like the expansion itself, was isotropic – the same in every direction – like the expansion. Some of the universes that Taub uncovered had the new feature that the curvature of their space could be different in each direction in space (Figure 7.6). The simplest anisotropic universe, found by Edward Kasner, has anisotropic expansion but at any moment of time space is flat and its curvature is isotropic. This turns out to be a special situation. The most complicated of Bianchi's spaces have both anisotropic expansion and anisotropic curvature of their space at each moment of time.

These new universes with anisotropic curvature look like the simple ones with isotropic curvature but with waves of gravitational energy added. The waves move in different directions, with different strengths, and their presence creates different curvature in different directions as they move. For all these complicated things to happen at once and still keep the universe looking the same from every place within it is very constraining and this is why there are just a small number of possibilities, the nine first found by Bianchi. If you allow things to be different at every place then the number of possible universes becomes infinite. Taub's group of universes is like the restriction an artist places upon his creativity by enclosing it in a frame or a sculptor who has chosen to express herself only in stone. Self-imposed restriction brings focus.

SMOOTH UNIVERSES AND A NEW OBSERVATIONAL WINDOW

It is always dark. Light only hides the darkness.
Daniel McKiernan

In the last chapter we learned about the remarkable serendipitous discovery of the cosmic microwave background radiation left over from the hot early stages of the universe. Its average temperature was deduced to be close to 2.7 degrees Kelvin but radio astronomers in Princeton soon realised that they could make a different type of measurement with far greater accuracy. It is very hard to measure the temperature with great precision because you need an absolute reference temperature against which to compare it. Today, the best temperature measurements are made in space, from satellites, but they have to carry an insulated flask of liquid helium at a temperature near 2.7 degrees to act as a comparison reference. Eventually, that liquid evaporates away and this limits how long the experiment can last for.

In 1967, Dave Wilkinson and Bruce Partridge exploited a clever piece of electronics invented by Robert Dicke to measure the possible differences in temperature between directions in the sky far more accurately than anyone could measure the average temperature. Measurements of the differences can be done very accurately because they don't require a measurement of the temperature. By searching for changes in intensity of the background radiation, and finding none to the level of sensitivity of their detector, Wilkinson and Partridge were able to show that temperature variations from one direction to another were less than 0.1 per cent: the background radiation was extraordinarily isotropic and there were no huge lumps of matter in the universe that were distorting its expansion significantly. [18]

This was a very striking discovery. It showed that the universe was expanding extremely isotropically; so isotropically, in fact, that cosmologists started to rethink their attitude towards the high degree of symmetry witnessed in the universe. Previously, cosmologists had *assumed* the universe was very nearly isotropic and homogeneous. The big puzzle was explaining why there were small irregularities on top of this smooth background and how these little irregularities grew to become the galaxies and clusters we see today. Soon after the high level of isotropy in the background radiation was discovered, cosmologists began to regard the smoothness of the background and the almost perfectly isotropic expansion as the major mysteries. After all, if we were picking universes at random from solutions to Einstein's equations or from Bianchi's gallery of spaces, it would seem far more likely that we would pick an irregular one that expands anisotropically than a beautifully smooth and isotropic one. There are just so many more ways for a universe to be irregular, so why is our universe so improbably smooth and symmetrical?

CHAOTIC UNIVERSES

'All roads lead to Rome.'
    Jean de la Fontaine[19]

Spurred by these observations of the microwave background radiation in 1967, Charles Misner at the University of Maryland advocated a completely new approach.[20] Instead of assuming the universe is smooth and isotropic, so order comes only out of order, why not try to prove that order will always result from chaos?

Misner's 'chaotic cosmology' programme sought to show that Einstein's universe equations have the property that, no matter how chaotically universes start out expanding, if you wait long enough (and we have nearly 14 billion years of history available so far) they will all become smooth and increasingly isotropic.

This is an appealing idea. It has major philosophical implications. If true, it would mean that we don't need to know how the universe began (or even *if* it began) in order to understand its present structure. Misner's idea was to show that if a universe begins in a chaotically expanding state, there are inevitable sources of friction arising in the early stages that smooth out the irregularities and ensure that the universe always ends up expanding uniformly and isotropically. It is like taking a bucket of oil and giving it a vigorous stir. You can close your eyes so you can't see the oily maelstrom you are creating. But you know that one minute later the oil is always going to look the same. The motions will have been damped out and its surface will always be smooth and placid. Could universes be the same?

This new chaotic cosmology was very ambitious. Previously, astronomers had been content to find one of Einstein's possible universes which described what we saw as accurately as possible. The simplest and most symmetrical candidates worked very well. Then the observations of the background radiation highlighted the fact that they did better than work well – they worked amazingly well. Why? That was the question that Misner raised. Some cosmologists were content to accept that the universe was highly symmetrical today because it had started out that way – 'Things are as they are because they were as they were', as Tommy Gold caricatured it.[21] That isn't really an explanation and Misner wanted to do better. If he could show that the present symmetry arose from any possible starting conditions (or even just most of them), then he would have provided a very satisfying explanation for the structure of the universe.

At first, Misner's approach looked quite promising. The simplest anisotropic universes based upon the types of space that Bianchi, Kasner and Taub had found did indeed tend to quickly become increasingly isotropic as the expansion proceeded. Unfortunately, the more complicated possibilities, which were the more likely, didn't share this nice property. You couldn't just rely on the expansion of the universe to make the irregularities less and less significant. In some universes they don't die away at all unless some specific physical process arises to downgrade them more quickly than the universe is expanding. The challenge was to seek out those smoothing processes

Unfortunately, this search was shown to be hopeless. Richard Matzner, of the University of Texas at Austin, and I showed that the background radiation itself was telling us that hardly any irregularities could have been erased in the past.[22] All frictional dissipation of irregularity generates heat – that is where the irregularity energy must end up thanks to the second law of thermodynamics – and the microwave background is where that heat resides today. The earlier the frictional processes operate, the more heat they produce. The total amount of heat in the universe today – about a billion photons for every atom – turns out to place a very restrictive limit on how much irregularity could have been damped down and eradicated in the past. In fact, only a very tiny amount of irregularity could have been dissipated in the early stages of the universe's history: it wasn't possible for universes that began in a very irregular state to be smoothed into the nice isotropic expansion we see today without impossible amounts of heat radiation being produced.

Misner also highlighted an unusual feature of all the universes that had been studied so far, whether they were highly symmetrical or not. Light travels at a finite speed and defines a cosmic speed limit for any type of signal or frictional process to operate. When the universe has aged by t seconds there has only been time for light to traverse a 'horizon' distance of about $10^5 \times t$ kilometres and that contains a mass equal to $100,000 \times t$ times the mass of the Sun, so after 10 seconds it can have travelled only a million kilometres (the distance from the Earth to the Sun is 150 million kilometres) and affected a mass of a million times that of the Sun. Yet we observe uniformity in the universe out to dimensions that are 10^{15} times bigger than this.

This conclusion cast a further huge shadow over the chaotic-cosmology philosophy. Any process able to damp out differences in temperature, density, and expansion rate from place to place in the early stages of the universe is limited to very small scales which cannot explain why the large-scale universe is so similar from place to place and why the background radiation temperature is so similar all around the sky.

If we look at different parts of the sky separated by more than two

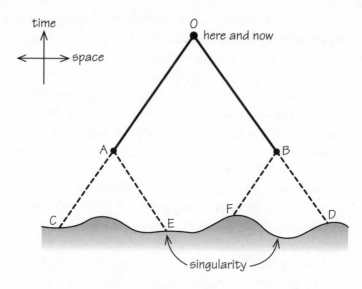

Figure 7.7 Light rays travel towards us through space and time down a 'light cone' shown in the diagram. It defines our visual horizon. Parts of the universe outside this past light cone OCD are observationally inaccessible to us. Take two points (A and B) on our light cone in the past where the photons of the microwave background first started to fly freely towards us, when the universe was a few million years old. We can find the past light cones (ACE and BFD) of those points. They do not intersect during the whole past history of the universe back to the beginning. This means that no light signal has had time to pass between A and B during the history of the universe: conditions at A and B cannot be coordinated by physical processes moving at the speed of light. Yet the temperature and density at A and B are found to be the same to an accuracy of one thousandth of one per cent.

degrees which have the same temperature to high precision, then there was not enough time for light signals to pass between them and even out differences in energy and temperature before the photons started flying freely towards us when the universe was about 250,000 years old. Our observations of the background radiation are snapshots of that moment.

This simple deduction meant that you either had to imagine some type of 'new physics', which could evade the cosmic speed limit on smoothing, or some special starting condition for the universe that rules out the exotic irregular universes as descriptions of the real cosmos.

MIXMASTER UNIVERSES

Harmonic mixing is an advanced technique used by top DJs all over the world . . . Obviously learning how to mix in key is an advanced DJ technique, if you're not advanced enough in your learning to start mixing in key just yet that is fine just make a mental note that mixing in key is a future step you are going to need to take to truly become a world class DJ.

The DJ Master Course[23]

In 1969, Misner responded to the limitations on the scope of dissipation in the early stages of the universe by finding a new type of solution to Einstein's equations.[24] Its space had one of the most complicated forms found by Bianchi and was finite in volume. It was a 'closed' universe that began at a big bang, reached an expansion maximum and then contracted back to a big crunch. However, the way in which it expanded was so intricate that it was not possible to solve Einstein's equations at all, although one could map out the overall pattern of behaviour and study parts of it by computer. Later, it was shown that its behaviour is chaotically unpredictable,[25] a feature that became a major area of interest to science in the late 1970s.

Misner called his new type of universe the 'Mixmaster' after an American commercial food mixer,[26] because he believed that its highly contorted, rapidly changing, geometry could allow light to travel round the universe and homogenise it. Unfortunately, detailed studies revealed that although light could travel a long way around a Mixmaster universe, this was a very rare occurrence and mixing was generally ineffective.

The Mixmaster universe was the most complicated one of Einstein's universes that had ever been seen (Figure 7.8). As you traced it backwards towards the beginning it underwent an infinite number of oscillations. At any moment, two directions expand and one contracts, just like in Kasner's universe. However, the contracting direction gets permuted around again and again, at random, and as you go backwards towards the beginning there will be an infinite number of permutations of the expanding and contracting directions. It's rather like a wobbling jelly.[27]

The reasons for this complexity are entirely to do with Einstein's picture of gravity: there is no counterpart of a universe like this in

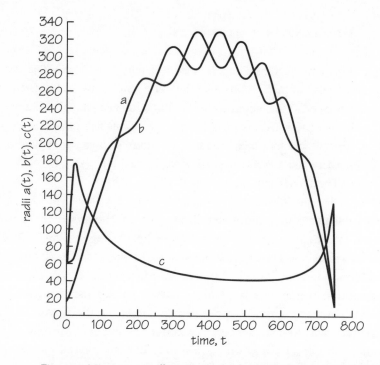

Figure 7.8 Mixmaster oscillations. For the first half of the universe's lifetime, its volume increases and two directions expand while the third contracts. During the second half, the total volume decreases to zero on approach to the big bang. The directions of expansion and contraction are permuted around in a chaotically random fashion, like a wobbling jelly, with periods in which two directions oscillate while the third changes steadily succeeded by new periods in which one of the oscillating axes is exchanged for the steadily changing axis. Only a few of the infinite number of oscillations are shown here.

Newton's old theory of gravity. You start with a universe that expands at different rates in different directions and then in each of those directions introduce gravitational waves which ripple through space in different directions. As they do so they curve the space in which they move. The wave moving in the imploding direction creates a large curvature which eventually turns the wave around, and the implosion reverses into expansion and one of the other directions now implodes. The sequence of changes is repeated infinitely often as you follow it backwards in time to the beginning.

This is an odd state of affairs that is linked to a very ancient philosophical question suggested by Zeno of Elea in the fifth century BC: can you do an infinite number of things in a finite time? Zeno liked to challenge other philosophers with beautifully conceived paradoxes of the infinite that were never answered in ancient times. For example, if the distance between you and the door is 1 metre, then to reach the door you must first go ½ metre, then ¼ metre, then ⅛ metre, then ¹/₁₆, and so on for ever. The total distance of this infinite number of steps[28] equals 1 but, Zeno argues, you will have to make an infinite number of steps to reach the door and you will never get there.

The Mixmaster universe performs an infinite number of physically distinct oscillations in a finite amount of the time that we measure on our clocks. Misner argued that if you reject Zeno's paradox because the infinite number of sub-divisions of the distance from you to the door are not physically distinct events, you must judge the Mixmaster universe to be infinitely old because an infinite number of distinct physical oscillations of space happen before $t = 0$ is reached.[29]

Physically, this seems strange. Mathematicians are used to such things though: the graph of $y = x^2\sin(1/x)$, if you could draw it (which would need an infinitely fine pencil), would possess an infinite number of oscillations on any small interval of x that includes the value $x = 0$ (Figure 7.9).

Cosmologists are nervous of taking this 'beginning' at $t = 0$ too literally. We expect that the quantum theory must start strongly affecting the whole universe when it is younger than 10^{-43} seconds in age. This is the earliest time that the Mixmaster oscillations could make physical sense. But the universe expands so quickly compared to the frequency of oscillations that even if those oscillations continued all the way up until the present, there would only be time for about a dozen of them to have happened. In a realistic universe the number of oscillations falls far short of the infinite number needed for light to travel repeatedly out to great distances early on. Almost all of the oscillations occur in the immediate neighbourhood of $t = 0$, just like in Figure 7.9. This infinity is very 'busy' near the beginning.

Despite the disappointment that the Mixmaster universe was not the key to understanding the present structure of the universe, it has remained of strong interest. It is still the most general and complicated behaviour ever found in solutions to Einstein's equations and it remains

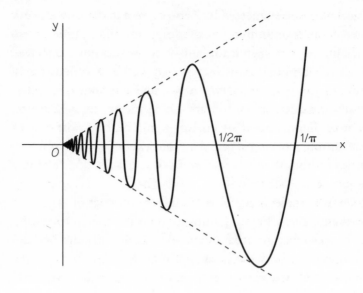

Figure 7.9 The graph of $y = x^2\sin(1/x)$. It experiences an infinite number of oscillations as it approaches $x = 0$. We are only able to draw some of them.

a challenge to see if it can be shown to be the most general type of universe that those equations permit.

Sadly, by 1980 arguments of this sort had persuaded cosmologists that the appealing idea of showing that physical processes in the very early history of the universe would always ensure the expansion became isotropic and homogeneous did not fly. There were just too many persistent types of irregularity to be got rid of and any process for smoothing was severely limited in scope by the finite speed of light and the amount of radiation entropy in the universe today.

MAGNETIC UNIVERSES

Examiner: What is electricity?
Candidate: Oh, Sir, I'm sure I have learnt what it is – I'm sure I *did* know – but I've forgotten.
Examiner: How very unfortunate. Only two people have ever known what Electricity is, the Author of Nature and yourself. Now one of the two has forgotten.
    Oxford University Natural Science viva, c.1890[30]

The sudden interest in universes that expanded at different rates in different directions led cosmologists to think about the types of energy that could only exist in such universes. Previously, they had focused entirely upon very simple possibilities: black body radiation and what they called 'dust' – a 'gas' of galaxies or stars which exerts no pressure (because the galaxies and stars don't often encounter each other). These are not the only energies and masses in the universe though. And one of the biggest long-standing mysteries has been the presence of magnetism.

Magnetic fields are all over the universe. They exist in planets, stars and galaxies. Where do they come from? Does the universe possess an all-pervading magnetic field from which all these smaller ones ultimately derive? One serious possibility was that they emerged from the beginnings of the universe, stretched and weakened by the expansion, along with all the matter and radiation. Later, they could be squeezed into galaxies by the force of gravity and be quickly amplified by rotation to produce the stronger magnetic fields that we see in and between the stars today. We still don't know if the origin of magnetic fields is as simple as this or whether some complicated process produced them, almost randomly, at a definite moment in the early history of the universe.

If magnetic fields were present right from the beginning then they require the universe to expand at different rates in different directions in order to sustain the anisotropic pressures and tensions that a universal cosmic magnetic field exerts. From 1965 to 1967, the American astrophysicist Kip Thorne and the Russian cosmologists Ya Zeldovich and Andre Doroshkevich found new solutions of Einstein's equations which described anisotropic universes containing ordinary matter or radiation together with a cosmic magnetic field. They confirmed the need for a very anisotropic beginning but they also revealed that the effect of the magnetic field on the later evolution of the universe is rather strange. It slows the fall-off of any anisotropy in the expansion very dramatically and leaves a distinctive impression in the microwave background radiation if its strength is too large. The absence of such a feature in the background radiation has enabled cosmologists to place very strong limits on the strength of any such universal magnetic field.[31]

THE UNIVERSES OF BRANS AND DICKE

'Don't believe an experimental result until it is predicted by
theory.'
    Anon

During the 1960s doubts started to emerge about whether Einstein's
theory of general relativity was correct after all. Observations of the
orbit of the planet Mercury were worryingly different to what Einstein
had predicted. Eventually, this problem was resolved by a more accurate understanding of the turbulent activity on the Sun's surface – and
hence of the Sun's true diameter and shape – but in the meantime
there appeared to be a big problem. Dicke had been in the forefront
of new ways of testing gravity theories using geological and paleon-
tological evidence in conjunction with astronomy and had made
measurements of the Sun's shape in 1966 that were problematic for
general relativity. The difficulties were not fully resolved by other
observations until 1973.[32]

Recalling the old work of Dirac on Large Numbers, and the possibility that Newton's gravitation 'constant', G, might not be constant,
in 1961 Dicke and his research student Carl Brans developed an impressive generalisation of Einstein's theory (Figure 7.10a).[33] They promoted
the gravitation constant from just being a constant to being a quantity,
like density or temperature, that could vary in space and time. If this
is done properly, rather than just by changing G into a variable in the
equations where it does not vary, as Dirac did, then a tightly constrained
new theory emerges. The variations are not arbitrary. They must
conserve energy and momentum, and like all other forms of energy
they will curve space and change the flow of time. If G is made to
be a constant then the theory of Brans and Dicke just turns into
Einstein's theory. But by allowing a very slow variation to occur, Brans
and Dicke's theory led to changes in the motion of the planet Mercury
so the observations (later to be shown to be incorrectly interpreted!)
could be explained.

The universes of Brans and Dicke were very similar to those in
Einstein's theory. Some expanded for ever, some reached a maximum
size and recontracted, but there were small differences. The gravita-

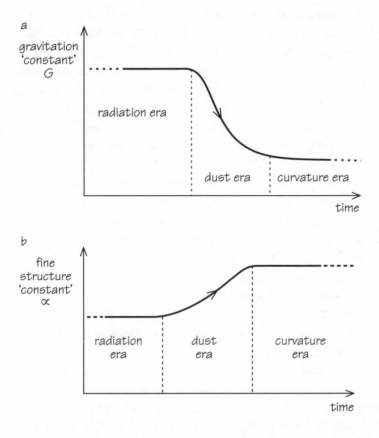

Figure 7.10 (a) Typical behaviour of a changing gravitational 'constant', G, over the age of the universe in cosmologies predicted by the gravity theory of Carl Brans and Robert Dicke. The value of G changes significantly only when pressureless matter ('dust') dominates the density of the universe. Changes are halted in the radiation-dominated era or when curvature dominates an open universe. (b) The expected change in the value of a time-varying fine structure constant, α, in the cosmological theory of Sandvik, Barrow and Magueijo. No change occurs during the radiation, or curvature-dominated phases of cosmic history. During the dust era, α should increase very slowly in proportion to the logarithm of the age of the universe.

tion constant could fall much more slowly than in Dirac's original scenario. If G fell with time like $1/t^n$ then the expansion of the universe increased in scale as $t^{(2-n)/3}$, so when n was zero and there was no change in G with time at all, we had the simple universe of Einstein and de Sitter with scale growing[34] as $t^{2/3}$ (Figure 7.10a).

A variety of studies of astronomical data and tests of gravity in the solar system revealed that Einstein's predictions of Mercury's motion were accurate and the value of n was pushed closer and closer to zero. By 1976, interest in these theories had waned but, twenty-five years later, they re-emerged as an important guide to create generalisations of Einstein's theory in which other traditional constants of Nature could vary and be self-consistently described.

Starting in 1999, there grew up an impressive body of astronomical evidence from the light spectra of quasars that a combination of the constants of Nature (like the electric charge carried by an electron, the velocity of light or the mass of an electron) which govern atomic physics may have been slightly different 10 billion years ago.[35] These observations are consistent with a very slow increase in the value of a combination of constants of Nature called the 'fine structure constant'[36] over recent cosmic history, at a rate that is a million times slower than the expansion rate of the universe. The novelty of these quasar observations, and the reason we first made them in 1999, was that they were more sensitive to changes in the fine structure constant than any laboratory experiment.

These observations needed a generalisation of Einstein's theory of gravity that could self-consistently include variations in the fine structure constant, which governs the strength of electricity and magnetism. A theory analogous to that of Brans and Dicke was found in two steps by Jacob Bekenstein[37] and Håvard Sandvik, João Magueijo and myself[38] in 2002 and provides a simple prediction as to how the constant should change in time (Figure 7.10b). Electricity and magnetism play no significant role over astronomical scales, so this theory produces no measurable changes to the expansion of the universe although the expansion determines the slow increase of the 'constant' with time.

MATTER—ANTIMATTER UNIVERSES

'Oh dear, what can the matter be?'
    English nursery rhyme[39]

The discovery of the microwave background in 1965 produced a sudden focus of interest upon the big-bang universes and the abandonment of traditional alternatives like the steady-state model. The mainstream arguments about 'which universe' we live in were about whether the hot early universe was isotropic, anisotropic or even totally chaotic. Yet there were still those who didn't like the big-bang picture at all, or who retained a philosophical penchant for the steady-state model because it had no beginning and no end: it was the most symmetrical universe possible because it was symmetrical in time as well as in space. There were others still who rejected both the big-bang model and the steady-state and pursued quite different pictures of what might have gone on in the very early history of an expanding universe. One interesting theory of this sort was motivated by the role of antimatter in the universe.

Following Paul Dirac's 1928 prediction that antimatter[40] should exist in Nature, it was discovered first in the form of anti-electrons ('positrons') from space by Carl Anderson in 1932; later, antiprotons were made and identified in 1955 by Owen Chamberlain, Emilio Segre, Clyde Wiegand and Tom Ypsilantis at the University of California at Berkeley.

Soon the cosmic significance of antimatter became a topic of debate. In the mid-1960s two leading Swedish physicists, Hannes Alfvén and Oskar Klein, proposed that the universe began with equal amounts of matter and antimatter in a greatly extended, low-density state – the complete opposite of the big bang. The matter and antimatter would be slowly drawn towards each other by gravity and eventually particles would start colliding with anti-particles, annihilate each other, and generate huge amounts of radiation. This blast would stop the infalling particles and anti-particles and drive them back out in a state of expansion. This might not happen everywhere but it was the expansion we see today, they argued, because we are necessarily in one of those regions where particle–anti-particle annihilation reversed the contraction into expansion.

Alfvén was not a cosmologist but a specialist in plasma physics and magnetic fields who won the Nobel Prize in physics for that work in 1970. The Alfvén matter–antimatter cosmology attracted little serious attention and was quickly shown to be at variance with a number of observations.[41] It required overall matter–antimatter symmetry in the universe: for every atom there was an anti-atom, for every star an anti-star. We didn't see any of those anti-atoms, anti-planets and anti-stars so the annihilation had to have led to widely separated regions of all matter and all antimatter – but how? Worse still, the reversal of contraction into expansion would have occurred when the density of the universe was 100 times less than we see it is today. This type of cosmic bounce could not have happened in our past.

Others, with deeper knowledge of high-energy physics, like Roland Omnes, tried in the late 1960s to develop a theory of the matter–antimatter universe in the context of a big-bang universe. Again, the initial assumption was that the universe contained equal amounts of matter and antimatter and that this symmetry was preserved; at the time, the unchangeability of the overall matter–antimatter inbalance was generally believed to be an inviolate law of Nature.[42] In order to reconcile this with the absence of any evidence for any antimatter in the universe near us, there had to be separated islands made of all matter and all antimatter around the universe.

Alas, if we run the expansion of the universe backwards in time those islands are going to be running into one another. They must have begun all mingled together with radiation in a state of equilibrium with particles and anti-particles appearing and disappearing into the sea of radiation during the first millisecond of the universe's history. It was possible to calculate what happened when equal amounts of matter and antimatter, democratically distributed, stopped annihilating each other. Unfortunately for this theory, the annihilation is very efficient and Zeldovich and Hong-Yee Chiu showed in 1965 that only one proton or antiproton survives for every 10^{18} photons.[43] What we see in the universe today are about 10^9 photons for every proton on the average. Our universe contains far too many protons and atoms and far too little heat radiation to have emerged from worlds of equal amounts of matter and antimatter in collision.

Antimatter was off the cosmological menu. But things were about to change. In the 1970s, a revolution was going to occur in cosmology that brought it back to centre stage and then ushered in the biggest of all shocks to our thinking about the universe.

8 The Beginning for Beginners

'To make an apple pie from scratch, you must first create the universe.'
   Carl Sagan

SINGULAR UNIVERSES

'Singularity is invariably a clue. The more featureless and commonplace . . .'
   Arthur Conan Doyle

Almost all the universes we have been describing had a striking feature. They 'began' at a finite time in our past where they had possessed infinite density. It was this singular beginning that had so repelled the creators of the steady-state universe and motivated them in their quest for an alternative that didn't spring into being at a special historical moment. Long before them, Richard Tolman had tried to avoid the inevitability of a beginning by imagining that the cycles of expansion and contraction of a closed universe were never-ending, like a bouncing ball. This required a suspension of disbelief at each moment when the universe bounced because Einstein's theory couldn't continue to hold true down there at such enormous densities and temperatures. You certainly couldn't continue the universe down to zero size and infinite density and expect everything to carry through unchanged.

When the first universes were found using Einstein's equations there were differing, but generally sceptical, views about the appearance of a beginning to the universe with infinite density. At first, Einstein thought that it was just a consequence of excluding pressure from the properties of the matter in the universe. If you included pressure, he thought that any attempt to run the universe backwards in time would eventually find that the pressure had grown large enough to stop further compression and the universe would rebound into expansion. It is like trying to squeeze a balloon into a smaller and smaller volume. The pressure resists this and eventually stops any further contraction.

Unfortunately, Einstein's intuition failed here. In his theory of gravity all forms of energy gravitate, and that includes pressure. Paradoxically, if you include pressure it doesn't stop the compression to zero size; it actually aids the contraction and makes the moment of infinite compression happen quicker: the gravitational effect of the pressure enhances the gravitational compression.

Next, Einstein wondered if the infinite density was just an artefact of considering universes that were expanding perfectly symmetrically. If you reversed this spherical expansion everything would be coincident simultaneously at one point in the past, but if the expansion wasn't perfectly spherical things would 'miss' one another when you ran the expansion backwards, and the infinite-density event would be defocused and avoided. It was, in Einstein's view,[1] 'illusory' and this view was shared by other leading cosmologists, like Robertson[2] and De Sitter,[3] in the early 1930s.

This simple intuition, gained from thinking about the world like Newton, was soon shown to be false. In 1932, Lemaître looked at a non-spherical, anisotropic universe and showed that it had an infinite beginning in the past just as surely as the isotropic ones.[4] The anisotropic universe of Kasner and Tolman's non-uniform universes all exhibited the same infinite big bang in the past.[5]. Still, perhaps there were more complicated asymmetries, like rotation, that could stop the universe having experienced infinite density in the past – an event that eventually became known as the initial 'singularity'.

This infinity is striking. If you walk around on the right parts of the Earth's surface you can find rocks that are over 4 billion years old; the simplest forms of bacterial life are about 3 billion years old and our modern human ancestors emerged about 200,000 years ago. The Earth and the solar system are not very much older than those surface rocks, about 4.6 billion years. Surprisingly, the expansion of the universe is suggesting that if we were to go back to a time just three times earlier – 13.8 billion years ago[6] – there would be no time, no universe, no *anything*. That is a stunning claim. We are seemingly quite close to the beginning of everything that is.

Despite the discovery that simple pressures and asymmetries could not exorcise the initial singularity, there was an influential view in the early 1960s that this singularity was not a physical feature of the universe models and so we should not worry about its physical reality. A group of Russian physicists, inspired by Landau and led by Evgeny Lifshitz, argued that the big-bang singularity, with its infinite density and beginning to time, was entirely fictitious and quite innocuous.[7] By way of a familiar analogy, think of a geographer's globe of the Earth. It will be covered with a network of lines of latitude and longitude, which are used to uniquely label each point on the Earth's surface. We call them 'co-ordinates' because they enable us to co-ordinate different places on the Earth. As we look towards the North and South Poles we see that the lines of longitude all converge and eventually intersect at the Poles: the map co-ordinates have degenerated into a 'singularity' of a particular sort. This doesn't mean that anything peculiar has occurred on the Earth's surface. We have simply chosen those co-ordinates to map the surface. We can always change to a different grid system near the Poles so that our map has no breakdown in its co-ordinate representation there at all.

This is what Lifshitz and his colleagues argued was happening as we followed our expanding universes backwards in time. The big bang was merely an innocuous mapping singularity created by using a bad choice of co-ordinates to describe the universe.[8] When we run into that fictitious singularity we should switch to a new description, and if that fails, to another still. The Russian group therefore concluded that the big-bang singularity was not physically real. It was no beginning of the universe. Unfortunately, it was eventually realised that this avoidance of the singularity by a sequence of map co-ordinate changes is also an illusion: it fails to ask what happens if you keep on making changes to the co-ordinates used. When that is carefully explored it is found that a true physical singularity remains, just as a real hole in the Earth's surface couldn't be removed by continually changing mapping co-ordinates.

This created a confusing situation in the early 1960s that needed cosmologists to think more carefully about what they meant by a

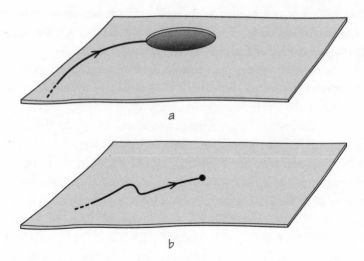

a

b

Figure 8.1 Two sheets representing universes where paths of light rays come to an end. In (a) there is a hole and the light ray hits its boundary. In (b) the light ray runs into a place with infinite material density where space-time is destroyed.

singularity. If we think of a universe model at any moment of time as a great sheet of space then we can locate singularities on that sheet by asking where the density becomes infinite. Now, suppose we just cut around those points and throw them out. We are left with a new (perforated) sheet of space which describes a perfectly good universe with no singularities. This seems like a bit of a cheat. The perforated universe is surely almost singular in some sense? But if we find a singularity-free universe how do we know that our method of finding and describing it hasn't inadvertently 'cut out' the real singularities in this rather artificial way?

An answer was proposed by Charles Misner in 1963 which subsequently changed the way in which cosmologists thought about singularities. [9] The idea was to give up the traditional notion of the singularity as a place where the density of matter, or some other physical quality, becomes infinite. Instead, we say that a singularity occurs if there is some path through space and time of any particle of matter or ray of light (its 'history') which comes to an end and cannot be continued any further.

What could be more 'singular' for a traveller moving along one of

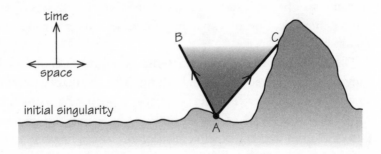

Figure 8.2 The initial singularity need not be simultaneous. The earlier beginning of the universe in some places, like A, might be visible by looking at light arriving from a place like C with a much later beginning. The lines AB and AC are paths of light rays from A.

these paths than this Alice-in-Wonderland experience? At the end of its path our light ray reaches the edge of space and time. It is no more.

The elegant feature of this simple way of defining a singularity is that if a physical attribute, like the density or energy of matter, becomes infinite somewhere then the light ray's path will be brought to a halt, because space and time will be annihilated there. But if this dangerous place had been inadvertently excised from the map of the universe then the light ray's history would still come to a stop when it reached the perimeter of the hole that remains after the excision (Figure 8.1).

This picture of the singularities as the edge of space and time is an extremely useful one. It bypasses those questions about the shape of the universe that had worried Einstein so much and it avoids the ambiguities introduced by changes in mapping co-ordinates. If a universe is non-singular, then it must be possible for all the possible histories followed by particles and light rays to be traced back indefinitely into the past: not one of them must have a past end-point. A non-singular universe is one that has no holes, edges or missing points.

Despite the simplicity of this idea it spawns lots of difficult questions. It might be that every time a past history comes to a full stop there is also an infinite density, energy or temperature that is responsible for destroying space and time, just as in the earlier intuitive idea of the big-bang beginning that we find in the Friedmann–Lemaître universes. But we don't know if that is generally true; even today, it remains an open question, although infinities in physical quantities are a general feature of expanding universes under a wide range of conditions.

Other changes to our commonsense expectations about the beginning of the universe are more striking. The singularity need not be universal: not all past histories need have a beginning. Some of them might have a start while others continue back into the past without end. And even if every possible history did have a beginning, there is no need for them all to begin simultaneously. Most dramatically of all, the beginnings of some histories could be observed by astronomers following other histories elsewhere (Figure 8.2).

WHICH UNIVERSES ARE SINGULAR?

'What a strange little object is the singularity with its strange properties and nonexistent definition . . . Here is a problem with which we must some day come to grips.'
    Robert Geroch

After all the confusions and uncertainties of deciding what a singularity was and whether it could be avoided, in the early 1960s cosmologists wanted to find a clear-cut way of deciding this question for universes in general rather than simply by examining each solution they found, one by one, to see if it had a beginning or a singularity in Misner's sense. An almost unnoticed start had been made on this by the Indian mathematician Amalkumar Raychaudhuri[10] working in Calcutta in 1953, and then by Arthur Komar[11] in America in 1956. Neither of them liked the idea of singularities and Raychaudhuri, perhaps without the traditions of a Christian heritage to influence him, was not looking for universes to have beginnings. Even if the density became infinite at some time in the past, he didn't see any need to equate this with the beginning of the universe. It might be possible to continue through the infinity in some as yet unknown way to an earlier state of the universe. More likely, he thought, was that Einstein's equations simply failed to give a complete description of the universe once the density became too large. New terms would appear in Einstein's equations; the solutions could change; and they might no longer have singularities.

This was also Einstein's attitude to singularities. He believed, just like Aristotle 2000 years before him, that there should be no physical infinities in the universe. Any that appeared in mathematical descriptions of the universe must be artefacts of an inappropriate idealisation in the description of the universe, or a breakdown in the assumptions on which the mathematical theory producing that model was based.

None the less, working with Einstein's equations, both Komar and Raychaudhuri showed that under very general conditions there would be times in the past when asymmetrical, but non-rotating, universes experienced conditions of infinite density just like the isotropic universes of Friedmann and Lemaître.

The exclusion of rotation was unfortunate because it could stop infinite densities developing in Newton's theory of gravity. But this lacuna along with the old ambiguities about mapping co-ordinates developing confusing singularities were all resolved by a sweeping development in 1965, which provided a new way of looking at universes.

Roger Penrose was a young pure mathematician, working on algebraic geometry, who had been tempted to think about Einstein's theory and the problem of singularities in universes by Dennis Sciama, a fellow research student and friend of Roger's older brother, Oliver. Sciama had worked with Hoyle, Bondi and Gold on aspects of the steady-state universe in the 1950s but had undergone a rapid conversion to the big-bang perspective after the discovery of the microwave background radiation in 1965.

Penrose used his familiarity with pure mathematical methods never before used in the study of universes to think about the question of singularities in a completely new way. Instead of trying to show that Einstein's equations led to past infinities in density, like Komar and Raychaudhuri, or inspecting one solution after another describing different universes, or using uncertain approximations like Lifshitz and his collaborators, Penrose changed tack. Using the definition of a singularity that Misner introduced, he focused on the equations that described the histories of particles and light rays. He proved that at least one of those histories had to have a beginning under very general conditions, irrespective of the details of the universe.

Penrose established the first 'singularity theorem' of this revolutionary sort in 1965, to establish that a singularity occurs inside a black hole if a massive dead star contracts under the force of its own gravity.[12] Then, in 1965 and 1966, Stephen Hawking, George Ellis and Robert Geroch used Penrose's methods to do the same for entire universes.[13]

A situation of considerable subtlety then slowly emerged as cosmologists tried to link up the old intuitive idea of the big bang, with infinities occurring in physical quantities, with the mathematical picture of a singularity as a history that had a beginning. Although they could reconcile the two in the simplest universes, like those of Friedmann, Lemaître, Kasner and Taub, to show that the unextendable past history was caused by the destruction of space and time at a *physical* infinity, this was not possible for all universes.

The spate of mathematical investigations eventually produced a very strong result, proved jointly by Hawking and Penrose in 1970.[14] Its attraction was that, despite being a difficult piece of mathematics, its assumptions could in principle be tested by astronomical observations. They showed that *if*:

1. space and time are sufficiently smooth (so we don't start out by inserting any glitches that cause histories to come to an end);
2. time travel is impossible (so that you can't avoid running into a beginning as you backtrack into the past by simply travelling around a time loop into the future);
3. there is enough matter and radiation in the universe (so that the force of its overall gravitational attraction is large enough);
4. gravity is always attractive;
5. and Einstein's theory of general relativity is always true,

then, there is at least one history in space and time that must have had a beginning.

Notice the logic here. This is a *theorem* not a *theory*. If those five assumptions are true then the conclusion is necessarily true. If one or more of the five assumptions fails[15] then that does not mean that a beginning can be avoided: it just means that we now have no theorem and the situation is indefinite. In fact, when some of these assumptions fail we find that we can have universes that are still singular or we can have ones that are non-singular, with no beginning.[16]

The reaction to these five conditions was varied. The first two were generally thought to be entirely reasonable – for most people the

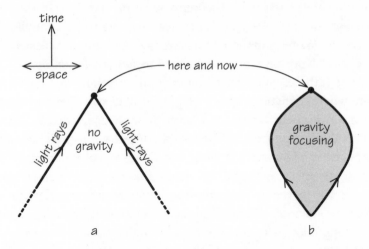

Figure 8.3 (a) Paths through space and time traced out by light rays moving at constant speed in the absence of gravity. (b) Gravitational attraction by the material and radiation in the universe bends the paths of the light rays from their straight-line paths. They are focused by gravity and will converge to a singularity in the past if there is enough material in the universe and if gravity is always attractive.

existence of time travel into the past would be worse than a beginning to the universe.

The third was something that could be checked by astronomers and shown to be true just by including the gravitational effect of the energy in the background radiation that had been recently discovered by Penzias and Wilson.

The fourth was believed to be true at that time. It was equivalent to the requirement that the pressure p and the density ρ of the matter in the universe at all times satisfied the condition that:

$$\rho + 3p/c^2 > 0 \quad (\star)$$

where c is the speed of light. For example, the black body or heat radiation that makes up the microwave background radiation in the universe – and is far away the predominant form of radiation in the universe – has the property that $\rho = 3p/c^2$, and because the density ρ is always positive, the condition $(\star)$ holds for the radiation and the effect of the gravitational field that it creates is attractive (Figure 8.3).

At the time when these theorems were proved, and all the way up until about 1977, this property (*) was believed to hold for all possible forms of matter and there was no reason to doubt it, unless you were the sort of person who was inclined to doubt everything.

The fifth condition was little different. It might well fail. But how and where? Newton's theory of gravity held sway for more than 200 years. Eventually, as we have seen, it was superseded by Einstein's general relativity theory. The two theories have considerable overlap though. When gravity is weak and things move much more slowly than the speed of light, Einstein's theory looks increasingly like Newton's in many respects. But Einstein's theory can go where Newton's cannot. When speeds approach that of light and gravitation is strong enough to force motions at light speed,[17] Einstein's theory still holds true and is internally consistent where Newton's fails.

If we look at different extreme situations we expect that Einstein's theory will ultimately fail as well. The discovery of the quantum nature of energy and matter by Bohr, Einstein, Planck and Heisenberg during the first quarter of the twentieth century revealed that energy comes in definite packets and cannot take all possible values. The traditional particles of physics are seen to have a complementary wavelike aspect. This wavy quality has more in common with a crime wave than a water wave. It is a wave of information. Just as the arrival of a crime wave in your neighbourhood means that it is more likely that a crime will be found to have been committed there, so an electron wave passing through your apparatus means you are more likely to detect an electron. The quantum wavelength of a mass m is inversely proportional to m so when the mass is large the effects of quantum waviness are tiny, they extend over distances far smaller than the physical size of the particle, and we don't have to worry too much about them. Yet when the mass of a particle is very small, the quantum wavelength becomes much larger and can easily exceed the physical size of the particle. In this situation the behaviour of the mass will look strange by Newton's standards: it will be intrinsically quantum – and wavelike.

Einstein's description of gravity does not include any of this quantum weirdness. We have therefore long appreciated, after John A. Wheeler stressed the point in 1957, that conditions could arise in the very early stages of the universe where Einstein's theory will fail. A new quantum

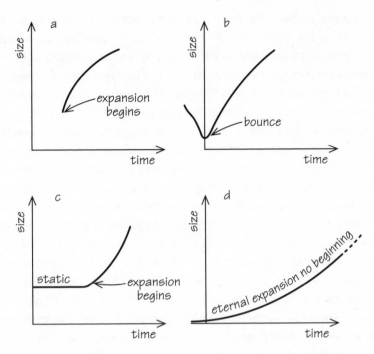

Figure 8.4 Some hypothetical pasts in which universes are non-singular and expand as observed today: (a) begins expanding at a finite past time but not at a singularity of infinite density; (b) starts expanding after a 'bounce' at an epoch of finite density following a phase of contraction; (c) expansion begins after an eternal past during which the universe was static; and (d) always expands, tracing an exponential curve which never reaches zero size at any finite past time.

version of general relativity – 'quantum gravity' – would be needed to take over. To find out when it is going to be needed we simply consider the distance light has travelled since the expansion began (the size of the 'visible universe') and ask when that is less than the quantum wavelength of the mass contained within a sphere of that radius (the mass of the whole visible universe then). Before that time, the whole universe will exhibit a quantum wavelike behaviour that Einstein's theory cannot describe: this new era of quantum gravity, when the very nature of space and time becomes uncertain, happens when the universe is less than 10^{-43} of a second old.

Suppose we follow our expanding universe backwards in time towards the type of historical beginning that the singularity theorem predicts and assume that it occurs everywhere simultaneously at a past moment of time that we will label '$t = 0$' for convenience. We know that we cannot trust Einstein's theory to give a full description of what occurs when we get very close to $t = 0$. We expect the theory to fail when we get back to $t = 10^{-43}$ seconds – a moment that has become known as the Planck time, after the physicist Max Planck.[18]. The failure could stop any theorem of this sort applying, either because space and time cease to be smooth enough or because the quantum gravitational effects required then bring about a failure of condition 4 in the theorem of Hawking and Penrose. Gravity could become attractive and cause the universe to 'bounce' back into a state of increasing size instead of continuing to contract back to ever-increasing density in the past. Alternatively, a range of simple alternatives can be imagined (Figure 8.4). All are consistent with the behaviour of the expanding universe we observe at later times. We can't choose between them.

This sounds reminiscent of the old cyclic-universe scenario that Richard Tolman proposed in the 1930s. Our singularity theorem means that the universe can't go all the way into the singularity and bounce out because space and time are expected to be destroyed at the singularity: you can't just go through it and close you eyes to the breakdown of Einstein's theory. The bounce has to occur when the universe reaches some very small, but non-zero, size and this requires condition 4 of the singularity theorem to fail.

Despite the likely failure of general relativity on approach to the singularity, from 1966 to 1972 cosmologists still regarded this theorem as predicting the inevitability of a past epoch of unimaginably high density and temperature. Even if a breakdown of Einstein's theory prevented a past big-bang event of infinite density, the density was still expected to have reached a value of 10^{94} times that of water, and that is pretty singular by any stretch of the imagination.

So far, we have talked about this singularity having occurred in the past, but it could also await us in the future if we inhabit one of those 'closed' universes first found by Friedmann whose expansion eventually reverses into contraction. As this final singularity is approached the universe will inevitably become very irregular because irregularities grow with the passage of time. Black holes will form and then be merged into one another. Some parts of the universe will experience a big crunch before others (Figure 8.5). Some observers may even

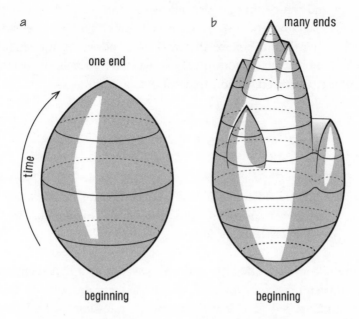

Figure 8.5 (a) Only in a perfectly smooth closed universe will the final singularity be simultaneous; (b) but in a realistic universe, where some regions are denser than others, the denser regions will hit the final singularity sooner than the sparser ones. An early encounter with the singularity in one place in the universe might be observable from other places.

be able to see other regions hitting the crunch long before them. Is a big crunch – the end of everything – occurring everywhere at different times? What does that mean – what does 'end' mean? What does 'everything' mean? It is not only the beginning of the universe that poses awkward and unanswerable questions.

The past singularity raises its own list of meta-scientific questions about what happened 'before' it. What determines what the universe is like when it emerges from the singularity? If space and time didn't exist before the singularity, how do we account for the laws of physics – did they exist before the singularity? How can you apply the usual methods of science to a unique event like the singularity?

These were some of the unusual questions that these studies of the beginning of the universe spawned. Cosmologists focused on more specific issues where answers looked possible. Could it be shown that the finite past histories always came to an end where there was infinite density and temperatures, just like when we extrapolate the model of our expanding universe backwards in time? The big question was whether we should believe that the assumptions of the singularity theorem hold true in Nature. If so, we have to believe that a past singularity occurred. If not, then all bets are off.

COLD AND TEPID UNIVERSES

'I know all your ways; you are neither hot nor cold. How I wish you were either hot or cold! But because you are lukewarm, neither hot nor cold, I will spit you out of my mouth.'
    The Revelation of St John[19]

The new conviction that our universe was one that had been fantastically hot and dense in the past, as the big-bang proponents had long been arguing, led to a new cosmological question in the late 1970s. There had been great emphasis upon the shape of the universe, how it became so smooth and isotropic, and whether it began in a state of chaos or order, but little attention had been paid to what it was made of. The discovery of the background heat radiation had created a new constant of Nature: the number of photons per proton (or atom) in the universe. This was a large number, roughly a billion.[20]

If we were to smooth out all the material in the universe, so that instead of stars and planets there is just an evenly spaced array of single atoms, there would be only about one atom in every cubic metre of space. This is an extraordinarily low density, far sparser than any artificial vacuum that could be made in a terrestrial laboratory. In the same volume there would typically be about a billion photons from the cosmic background radiation. So the ratio of roughly a billion to one is a measure of the relative heat radiation content of the universe. The value of a billion is very large, far larger than can be produced by exploding stars or other violent processes going on in the universe today, and was one of the reasons why it proved so difficult to find an explanation for the existence of this heat radiation in the steady-state universe. Still, the big-bang theory was left to explain why there were about a billion photons for every atom, and not vastly more or vastly less.

It was soon appreciated that this number played a crucial role in the history of a hot big-bang universe. It determined when conditions cooled off sufficiently for atoms and then stars and galaxies to form, and how big they would be when they did. But why there were a billion photons per atom remained a mystery. There were attempts to see if we could begin with big-bang universes that were 'cold', with just a few photons for every proton, or 'tepid', with about 10,000 photons per proton, and see if explosive cosmic events could boost those numbers naturally up to a billion.[21]. Unfortunately, these attempts didn't succeed. Like the turbulent universes before them, they couldn't provide an explanation for the amount of helium in the universe – tending to produce far too much – and all the violent events needed to generate the heat radiation would have left tell-tale variations in the background radiation temperature in excess of the one part in a thousand limit that had been established by astronomers. Hot universes needed to start out hot.

AN UNEXPECTEDLY SIMPLER UNIVERSE

All through the five acts he played the King as though under momentary apprehension that someone else was about to play the Ace.

Eugene Field, theatre critic[22]

One of the by-products of these investigations was to ask more searching questions about the elementary particles that populated the very early stages of the big-bang universe. It was a question that Lemaître had first raised but had been unable to answer. Then Alpher and Herman with James Folin had taken the first steps towards importing what we knew about high-temperature physics into the big bang universe.[23]

You might have thought that cosmologists just needed to team up with nuclear and particle physicists in order to find out what the universe must have looked like in its youth. Alas, before 1973 the high-energy physicists could be of surprisingly little help. The reason was a simple one. There was no working description of what happened to matter above 100 billion degrees, which was far in excess of temperatures attainable in any experiment on Earth. Unfortunately, the theorists were doing little better. All their attempts to create a theory of strongly interacting elementary particles ended up predicting that their interactions should just keep getting stronger and stronger as the energies increased. Very soon you had a completely intractable mess and no way of making reliable calculations.[24] This looked like the death-knell for understanding the very early stages of the universe. It was doomed to become increasingly complicated and intractable as we looked back to earlier and earlier times. The so called 'beginning' of the universe, or the singularity of Hawking and Penrose's theorem, seemed beyond the reach of our understanding of the laws of Nature.

In 1973, a path-breaking new theory of the behaviour of elementary particles at high energy emerged, owing to the work of David Politzer, David Gross, Frank Wilczek, Tony Zee and Gerardus t'Hooft. It turned our old picture of what happens in interactions between elementary particles of matter at high energies upon its head. The sea of energy in quantum vacuum all around us has an effect upon the qualities of the elementary particles moving within it. As a consequence, the strength of the interactions between the particles changes as the temperature of the environment increases. Put an electrically charged particle, like an electron, in a quantum vacuum and it will attract the oppositely charged particles from those that fleetingly pop in and out of existence on borrowed energy from the quantum vacuum. A negatively charged particle then gets surrounded by a shield of opposite charges. Another electron approaching it slowly, with low energy, as shown in Figure 8.6a, won't get repelled by the full negative charge of the electron but only by the less negative, shielded charge. On the other hand, a more energetic electron, as shown in Figure 8.6b, will penetrate deep into the shielding cloud and feel more of the bare, unshielded negative charge within it.

What this means is that the strength of the interaction actually felt depends upon the energy of the environment in which it is taking place. For particles like electrons, participating in electromagnetic interactions, the effective strength of the force they feel grows as the energy of the environment increases because of the quantum vacuum. For the case of the strong interaction between elementary particles like quarks and gluons, the effect of the quantum vacuum is totally unexpected: it makes the strong interactions get *weaker* as the energy

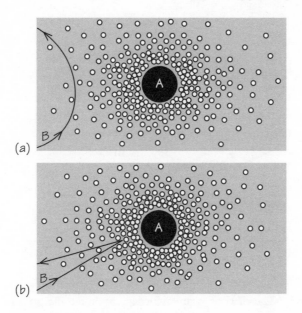

Figure 8.6 (a) An incoming electron, B, which has low energy, scatters weakly owing to a shield of virtual positive electric charges around the central negative charge of the target electron, A. (b) An incoming electron with high energy penetrates through the cloud of positive charges and feels a stronger unshielded force of repulsion from the target electron.

of the environment increases. This Nobel Prize-winning discovery is known as 'asymptotic freedom' because it suggests that at the highest possible energies particles would behave as if they were free of any interactions at all. It also served to free cosmologists from the shackles of old ways of thinking about elementary particles. Suddenly the early universe looked simpler.

AND A UNIFIED WAY OF THINKING

'One is one and all alone
And evermore shall be so.'
    'Green Grow the Rushes O'[25]

It took just a few years for these dramatic theoretical discoveries to alter the direction of particle physics. The change to the effective

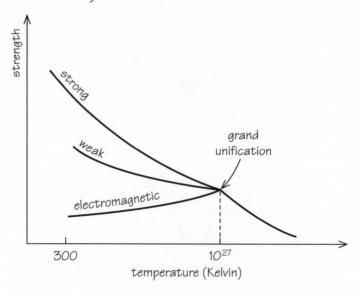

Figure 8.7 The effect of vacuum fluctuations on the effective strength of the electromagnetic, weak and strong forces of Nature predicts that the strengths will converge at very high energy and a 'grand unification' of these three forces of Nature can occur. At low energies in the universe today they have very different strengths.

strength of interactions between particles as energy and temperature increased offered a new solution to an old puzzle: how could there ever be a unified theory of all the forces of Nature when their strengths are so dissimilar? Superficially, strong and weak forces looked totally different. Not only did they have unequal strengths but they acted on separate populations of particles.

The increase in the effective strength of the weaker forces, like electromagnetism and radioactivity, together with the weakening of the strong force, could produce a triple cross-over at one very high energy where the effective strengths of the forces of electromagnetism and radioactivity and the strong force between sub-nuclear particles will all be equal. The idea of a 'Grand Unified Theory' (or GUT, as it became known) of these three fundamental forces was born when, in 1975, Howard Georgi and Sheldon Glashow found the first and simplest candidate for such a theory (Figure 8.7).[26]

All this changed the perspective of cosmologists and set the study of the universe in a quite different direction. All of a sudden the question of understanding how matter and energy were behaving at earlier and earlier moments in our reconstruction of the universe's past was no longer a hopelessly intractable problem, doomed to get worse the earlier into the past we probed. Asymptotic freedom meant that inter-action would get weaker and things become simpler (or, at least, no harder) as we looked back into the high temperatures that should have existed near the start of the universe's expansion history. As a result, the study of universes started to change. Previously there had been great emphasis on the geometry of universes, looking at all the different possible types of expansion, exploring whether they could evolve to become like the universe we see today, and trying to understand how the universe had emerged from a past singularity. Particle physicists sometimes took a passing interest in these discussions, but few were seriously involved in them. Once asymptotic freedom made it possible to say believable things about how matter behaved at very high tempera-tures, mainstream particle physicists converged on cosmology in growing numbers. The environment of the early universe offered a new place where they could work out the consequences of their theo-ries and seek out their observational consequences.

The interplay between the study of the universe – the very large – and the most elementary particles of matter – the very small – worked both ways. Particle physicists wanted to test out their new theories to see if they had good or bad astronomical consequences. Cosmologists started to take an interest in the unusual properties of theories of high-energy physics to see if they could shed light on some vexing problems about the universe. Could non-luminous matter in space be composed of new types of elementary particle? Why is the observed universe made of matter rather than antimatter?

Questions like these could be readdressed in the light of a new understanding of matter in the hot early universe. With it came a realisation that some quantities, like the cosmic balance between matter and antimatter, are not unchangeably fixed. And out of that new spirit of inquiry would emerge a host of new universes with surprises for particle physicist and astronomer alike.

9 Brave New Worlds

Imaginary universes are so much more beautiful than this
stupidly constructed 'real' one.
  G. H. Hardy

ASYMMETRIC UNIVERSES

The researches of many commentators have already thrown
much darkness on this subject, and it is probable that, if they
continue, we shall soon know nothing at all about it.
  Mark Twain

The entry of particle physicists into cosmology during the second
half of the 1970s led to a focus on using the new GUTs to solve the
problem of the matter–antimatter asymmetry of the universe and the
mystery of why there are about a billion photons for every proton in
the universe. These theories showed how a unified picture of the three
non-gravitational forces of Nature was possible even though each
appeared to be so different in strength. As the temperature of the
environment increases, their effective strengths change at different
rates and in different senses – the weak forces get stronger and the
strong forces get weaker – and they all converge on the same strength
at a very high energy that would be experienced when the universe
was a mere 10^{-35} seconds old.

This convergence also provided a solution to the other problem
about 'unification'. Existing theories which treated each force of
Nature separately did not allow all particles to interact with all others.
Quarks could not change into electrons, or vice versa. This seems an
unsatisfactory compartmentalisation of the elementary-particle world

into separate communities that don't converse across community boundaries. These separate communities were defined by whether the particles possessed certain attributes, like electric charge, or a more complicated version of it carried by the strongly interacting particles called 'colour charge'. Grand Unified Theories predicted the existence of new types of elementary particle which would carry both types of charge and could mediate exchanges between the previously separate communities of particle. These mediator particles would be very heavy and would only be produced profusely close to the energy where the effective strengths of the forces become similar. They were dubbed 'X' particles and they allow a true particle democracy to develop between different types of elementary particle in the very early stages of the universe.

These new interactions between particles had two immediate consequences that caught physicists' attention. The possibility of changing colour-charged particles, like quarks, into colour-free particles like electrons and neutrinos meant that the three quarks inside a single proton could decay. The proton was expected to be unstable.

At first, it looked as if the average lifetime might be small enough, 10^{30} years, to be seen in deep underground experiments where thousands of tons of material could be shielded from cosmic rays and other outside influences. Although the *average* lifetime of a single proton is vastly longer than the age of the universe (14×10^9 years), there are about 10^{30} protons in just 1000 kilograms of rock or water and by increasing the mass a thousand-fold we have a good chance of seeing the tell-tale decays of protons if we surround a large mass with sensitive detectors.[1]

Great excitement was created by early claims to have seen these decays in an experiment below the Earth's surface in the Kolar gold field in India that had run since 1980, but it turned out to be a false alarm.[2] Gradually, experimental limits have forced the lifetime of any proton decay to be much longer than these first optimistic expectations (greater than 6.6×10^{33} years[3]). Although the proton will be unstable in any unified theory, the lifetime is most likely to be far too long for us to detect or to distinguish from confusing 'mock proton decay' processes occurring in cosmic ray events.[4]

The second consequence of the new interactions was that quarks and their anti-particles, antiquarks, will not decay at the same rate.

This property, coupled with the decay of quarks into electrons and neutrinos, meant that the matter–antimatter balance in a universe could change. Now there was a possibility to calculate it and the answer needn't depend on what its value was when the universe began.

It depended on three things:[5] the fraction of the universe in the form of the particles that could provide the quark–antiquark-changing decays, the difference between the decay rates of the quarks and the antiquarks, and rate of the decays. The last factor enters because it was necessary for the decays to occur faster than the expansion rate of the universe in order to block the exact inverse reactions just restoring the matter–antimatter balance back to its starting value.

There were many attempts to do these calculations, increasing the complexity and detail of what was included.[6] The conclusion of them all was unanimous: it was indeed possible to explain the observed asymmetry between matter and antimatter in the universe no matter how it started out. Moreover, if the asymmetry between the quarks and antiquarks corresponded to, say, a billion and one photons for every proton, then when the universe cooled down to about 10^{13} degrees the billion protons would annihilate with the billion antiprotons leaving two billion annihilation photons for every unpaired proton left over. As a result, we end up with a universe containing about 2 billion photons for every proton,[7] pretty much as observed. The matter–antimatter balance and ratio of photons to atoms are intimately linked and can be explained together.

The importance of these studies was not so much that they identified the particular GUT of physics or the exact way in which the matter–antimatter asymmetry had been created in our universe: it was in showing that it was easily possible to provide such an explanation from a wide range of theories. When the right unified theory was identified it would be possible to predict or explain the matter–antimatter imbalance and the numbers of photons per proton. They were no longer just numbers that are as they are because they were as they were.

PROBLEM UNIVERSES

Never glad confident morning again!
　Robert Browning[8]

It didn't take very long before a profound cosmological problem raised its ugly head to threaten the whole idea of a Grand Unified Theory in physics[9]. If the weak, strong and electromagnetic forces were all one at very high energies near the beginning of the universe's expansion, then there was an unavoidable unwanted by-product. If the force of electricity and magnetism is to emerge from the early universe it will be accompanied by a profuse production of a new and very heavy particle, called a magnetic monopole, whose existence was first proposed by Paul Dirac in 1931.[10] It is like a very heavy (10^{20} times heavier) version of the electron, except that instead of an electric charge it carries a new type of 'magnetic charge', as if it is one magnetic pole of a bar magnet. These magnetic charges are conserved quantities in Nature, coming in plus and minus varieties just like electric charges, and the only way that a monopole can be removed is by hitting its anti-particle and annihilating. Unfortunately, once formed, monopoles hardly ever run into antimonopoles and the universe seemed to be stuck with them – lots of them.

This was a disaster. The universe today would be overwhelmed by magnetic monopoles. They would contribute 10^{26} times more to the density of the universe than all the stars and galaxies. In such a universe there would be no stars and galaxies, and no 'us' to worry about it either. This was the 'monopole problem'.[11]

The attempt to graft the new insights of high-energy physics into the reconstruction of the very early history of the universe was stopped in its tracks by this conundrum.[12] Strangely, for the first time, the study of universes found itself being steered by theories about elementary particles of matter rather than by new astronomical observations or solutions to Einstein's equations.

INFLATIONARY UNIVERSES

The baby figure of the giant mass
Of things to come at large.
    William Shakespeare[13]

These problems with monopoles caused something of an impasse but they kept the focus on universes as vast theoretical laboratories for particle physics as opposed to expanding geometries of stars and galaxies. Particle physicists were only interested in assuming the simplest possible type of expansion: a universe with smooth and isotropic space, with no rotation or other unusual features at all.[14]

Suddenly, in 1980, a new typescript appeared in physics department mailboxes. It was a substantial piece, written by a young particle physicist working at the Stanford Linear Accelerator Center. He was little known at the time and had been working in a succession of temporary post-doctoral assistant's positions in the USA in the midst of a major downturn in the academic job market.[15] Alan Guth had realised that there was a type of expanding universe that simultaneously offered a solution for several of the big problems that cosmologists had been wrestling with during the previous twenty years. This had not been his initial motivation, though: he had simply been trying to solve the monopole problem.

Guth's proposal, soon seized upon by many other physicists and cosmologists, was in essence very simple. It suggested that during the very early history of the universe, just before the events that may have produced magnetic monopoles and before the matter–antimatter asymmetry was established, the expansion of the universe underwent a brief surge of *acceleration*, or *inflation*, as Guth called it (Figure 9.1) – ironically echoing the economic climate that had made it so hard to get a long-term job.

This was an unusual idea. We have already seen that there are accelerating universes like those first found by de Sitter and deployed in the steady-state universe. These universes *always* accelerate – from past to future eternity. But we have also seen universes, like Lemaître's, that would begin by decelerating but then eventually change gear and accelerate when the repulsive cosmological constant became stronger than the attractive force of ordinary gravity. Yet once they started accelerating they never stopped.

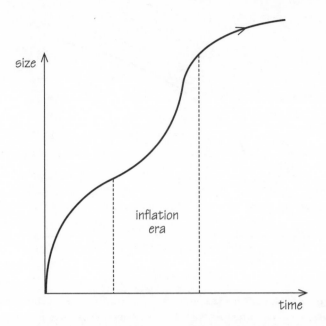

Figure 9.1 The distance versus time evolution of a simple inflationary universe. Inflation occurs if the expansion undergoes a finite period of accelerated expansion. This makes the universe expand far more, and stay much closer to the critical expansion rate for longer than if inflation had not occurred.

No one had found ever-expanding universes where acceleration, once started, could change back to deceleration (Figure 9.1). Guth had found a way to create a stress that was *temporarily* gravitationally repulsive, behaving for just a short while like Einstein's infamous cosmological constant.

The new unified theories of the behaviour of matter at very high energy contained new types of particle that eventually became known as 'scalar fields'.[16] These forms of energy can change very slowly, far more slowly than the universe was expanding, and if they did so they would exert gravitational repulsion rather than attraction upon each other. In the extreme and most common case they were almost exactly like a manifestation of Einstein's cosmological constant. They exerted a negative pressure, or tension, in the universe but unlike the pure cosmological constant they were transient: the scalar fields could just decay away, quickly or slowly, into ordinary radiation and other

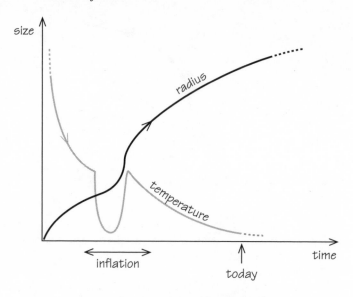

Figure 9.2 The surge in expansion scale during inflation is accompanied by a very rapid fall in temperature. When inflation ends there is a surge in energy created by the decay of the particles driving the inflation. This leads to a reheating of the universe. Subsequently the universe cools off at the same rate as it did before inflation occurred. Inflation makes the universe more expanded at any later time than if inflation had not occurred.

elementary particles that exerted positive pressure and were gravitationally attractive. The overall evolution of the size and temperature of the inflating universe is shown in Figure 9.2.

The new particle physics had offered some clues as to how this might be possible. The cosmological constant was, in the language of the particle physicists, a description of the 'vacuum energy' of the universe – the lowest energy that it could have. By this they meant a local minimum energy state, just like the position a marble will roll down to if you drop it into a curved kitchen bowl. Collections of elementary particles can choose from several of these temporary resting places, each with different energy levels, and it was possible to be moved from one to another by random jiggling by other particles as the universe cooled down (Figure 9.3).

If some of the matter in the universe found itself displaced from its equilibrium state, and started moving towards a new one with lower energy, it would experience the gravitationally repulsive stress

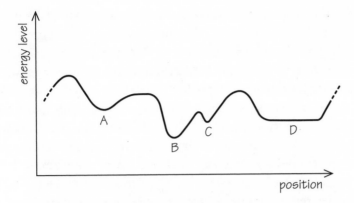

Figure 9.3 The early universe can contain matter that has a number of local minimum energy states, or 'vacua', which have different energies. In this picture they are located at A, B, C and D. The vacuum state B is the lowest of these and it might be possible for a part of the universe that finds itself lodged in the minima at A or C to receive a perturbation that enables it to jump over the 'hill' and fall down into B.

of the vacuum energy liberated by the move to the new lower energy state and rapidly accelerate.[17] Guth realised that the consequences of this change happening quickly might be disastrous. The changes would emerge like bubbles of new vacuum nucleating in different places at the same time. The bubbles would expand quickly and run into other bubbles. The result: a big mess of irregularities and huge variations in density and expansion rate from one side of the universe to another. This is not the universe we live in.

None the less, Guth put forward his new inflationary theory, high-lighting this problem, because it had a number of other remarkably simple and desirable consequences. Soon, others would recognise that it was possible for the transition from one vacuum state of the universe to another to proceed very slowly and for our whole visible universe to be inside one bubble of new vacuum. This meant that there would be no bubble collisions and accompanying debris inside our visible universe. If the right type of gravitationally repulsive form of matter appeared very early in the universe it could drive this acceleration but soon decay away into ordinary forms of matter and radiation. The acceleration would end and the universe would resume its decelerating course.

If the universe underwent this brief interlude of accelerated expansion there were several remarkable consequences that seemed to resolve some old problems about the universe. The universe grows much bigger, more quickly, than it otherwise would have. In the process, the expansion is driven very close to the divide that separates ever-expanding universes from those that eventually collapse back upon themselves towards a big crunch. At the same time, the universe becomes very smooth and increasingly similar in every direction in space (Figure 9.4). These were all previously unexplained features of the visible universe. The short period of inflation provides a natural explanation for all of them at once.

The most interesting consequence of inflation is that the period of faster accelerated expansion enables the whole of the visible part of the universe today (extending more than 14 billion light years across) to have expanded from a far smaller primordial fluctuation of mass and energy than was previously imagined to be possible – a fluctuation small enough to be kept smooth by light rays moving from one side to another.[18] The high level of smoothness we observe in our universe is therefore just a reflection of the fact that in an inflationary universe the whole visible universe is the expanded image of a tiny fluctuation that is kept smooth and isotropic by photons of light moving excess energy from the hotter parts to the cooler parts (Figure 9.5).

Last, but not least, inflation also solves the monopole problem. Monopoles are created because there can arise mismatches between the directions in which the magnetic stresses are pointing. In an inflationary universe the region from which our visible universe expanded is so small that none of these magnetic mismatches arise there and we don't expect there to be any monopoles. By contrast, if the universe had not accelerated early on, the region which expanded to become our visible universe would have to be at least 10^{25} times larger (about 1 cm across). There would be an enormous number of magnetic mismatches inside it, and a catastrophically large number of monopoles as a result, because the smoothing effects of light rays would be effective only over distances of about 10^{-25} cm.

Inflation provides a natural simultaneous explanation for the puzzling isotropy and smoothness of the visible universe, its avoidance of a glut of monopoles, and its proximity to the critical dividing line separating universes that expand for ever from those that will eventually contract to high density again. It has many interesting links to problems and universes we have already met. It has aspects that are

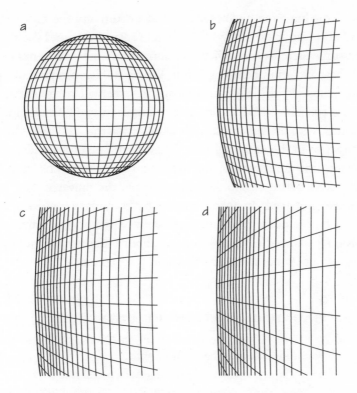

Figure 9.4 Progressive inflation of part of a curved space makes it appear locally flat as we go from (a) to (d).

reminiscent of the steady-state universe, but its 'steady-state' period is temporary. This is very important. If inflation doesn't end, and acceleration continues for ever, then everything gets smoothed out indefinitely. There will be no galaxies and stars.

Inflation also meets some of the challenges faced by the chaotic cosmology programme. Inflation succeeded in explaining the high degree of isotropy and uniformity without the need for any dissipation of irregularity and the problems that created. Instead of eradicating uniformities, inflation just swept them beyond the visible horizon in the universe today. They will still be there somewhere far away but the whole of our visible universe reflects the high isotropy and smoothness of a tiny patch of space that underwent inflation.

The earlier attempts by Misner and the 'chaotic cosmologists' to show that anisotropies in the initial expansion of the universe could all be damped out quickly enough to avoid messing up the good predictions of the big-bang nucleosynthesis of helium and the high level of temperature isotropy in the background radiation around the sky had never succeeded. Too much initial anisotropy and you generate too much heat radiation damping it out. Worse still, some types of anisotropy – those in the curvature of space – seemed to be stubbornly immune to being subdued in such a simple way. Cosmologists had been stuck with the very reasonable assumption that gravity will always be attractive in the early universe. They were open to the idea that it might become repulsive very late in the universe's history because there might be a positive cosmological constant in the universe, as Einstein had once proposed. In 1980, that wasn't thought very likely but it couldn't be excluded. Once the possibility of temporary gravitational repulsion is allowed early on, all the stubborn anisotropies get evened out very quickly because the expansion goes so quickly.

CHAOTIC INFLATIONARY UNIVERSES

If you can look into the seeds of time,
And say which grain will grow and which will not
    William Shakespeare[19]

The inflationary cosmology is content for the universe to begin in a chaotic and complicated fashion, just as the chaotic cosmologists believed was most likely (because there are so many more ways to be complicated than to be simple), but any tiny part of the chaos that is able to inflate will become larger, smoother and more isotropic, quickly pushing any other inflating regions far out of sight. Notice one very important point. The inflationary theory tells us nothing about the universe as a whole. We don't know whether it is smooth everywhere or totally chaotic globally, beyond our visible horizon.

But this tiny region that grows to encompass our entire visible universe will not be perfectly smooth. There will inevitably be very small statistical and quantum fluctuations which will end up being inflated in scale to seed the large-scale variations in density that we observe in the universe today in the form of galaxies. Without inflation to stretch them out, those statistical fluctuations would be initially too weak to explain the existence of galaxies by the process of gravitational

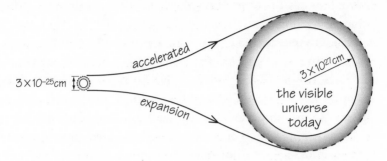

Figure 9.5 Inflation grows a region that is larger than the visible universe today (3 x 10^{27} cm) from the expansion of a patch of space small enough to be smoothed by light signals crossing it at any time in the early universe. In this picture we chose that early time to be 10^{-35} seconds; so light travelling at a speed of 3 x 10^{10} cm/sec will then be able to traverse regions about 3 x 10^{-25} cm in size. The temperature of the universe is 3 x10^{28} K at this time and falls to 3 K today. The overall temperature change factor of 10^{28} is the amount by which inflationary expansion stretches the region from 3 x 10^{-25} cm to 3 x 10^{27} cm. This explains the smoothness of the visible universe: we see the expanded image of a very small smooth region. If inflation had not occurred the visible universe would have expanded from a patch about 3 cm across – 10^{25} times bigger than the distance light rays could cross at that very early time. It could not have been kept smooth.

instability. All this became clear to cosmologists during a special two-week-long workshop on the new theory, held in Cambridge, during June and July 1982.[20]

It quickly became clear to the participants that inflation produced fluctuations with a particular pattern of variation. The pattern had been assumed for convenience in the past because it was the simplest one to adopt, with clustering having the same strength over every scale in the universe when it became visible.[21]. It was special. It was the only pattern of irregularity for which the universe would always remain looking like a Friedmann–Lemaître universe plus very small disturbances at all times in the past and in the future.

Using the fact that inflation just creates a temporary version of the steady-state universe one can see why it arises. Remember that the steady-state theory didn't allow you to distinguish the future from the past and de Sitter's exponential inflation guarantees that. If you have a natural process creating small irregularities, then you mustn't be able to use them to distinguish the future from the past either. If every size of fluctuation has the same strength when an observer sees it, then that principle remains unscathed.

So, as an unexpected bonus, inflation offered a possible explanation for the existence of galaxies and a particular pattern of small irregularities in the density of the universe. This would produce a special angular variation in the temperature of the cosmic background radiation on the sky. So, we now had a test to see if inflation had really happened.

Cosmologists have been looking keenly with an array of instruments for this 'smoking gun' from inflation and the evidence is building impressively to confirm it. The COBE and WMAP satellites flown by NASA have searched for the distinctive pattern of variations expected in the temperature of the incoming cosmic background radiation as we compare its temperature in different directions in the sky. Satellites do better than Earth-based instruments because they do not have to look through our changing atmosphere and they can scan the entire sky, building up a huge number of temperature comparisons so that the purely chance variations that arise in any data set are rendered insignificant.

These observations and the theoretical predictions they are testing have been represented to show the strength of the temperature differences against the angular separation on the sky over which they are measured. The predictions of the standard inflationary-universe model over a range of angles are shown with the solid line in Figure 9.6. There are distinctive features. Oscillations decay away towards small angles, like the ringing of a bell. As we go to the right we are probing smaller and smaller scales and eventually all the fluctuations will be ironed out by physical processes that transfer energy from over-dense regions to under-dense ones. If we extended the solid line to the left then it would go horizontal and be in very good agreement with the observations taken by the old COBE satellite, which were confined to separations on the sky exceeding 10 degrees. The data points which accurately trace out the solid curve come primarily from the WMAP satellite mission.[22]. We notice that there is a very close agreement with the predicted inflationary pattern over the first few bumps but then the observational uncertainties become large as we reach the limits of the instrument sensitivity. There is a peculiar 'dip' in the signal as it straightens out on large angular scales which has been a subject of considerable argument and interpretation among astronomers.[23]

A mission by a new satellite, called *Planck*, launched in the summer

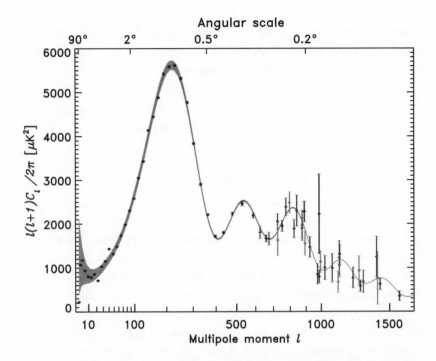

Figure 9.6 The temperature fluctuation level (in units of micro Kelvin squared) predicted by an inflationary universe (solid curve) against angular scale on the sky in degrees compared with the observational data from various ground-based and balloon experiments. For comparison, the full moon extends about 0.5 degree. The principal data on large angular separations is from the WMAP satellite after seven years of data-gathering (black dots). The shaded band on the curve towards the right is the irreducible statistical uncertainty created by the limited number of regions of this size in the visible universe.

of 2009 by the European Space Agency, will soon improve the accuracy of the observations on small angles even further. Meanwhile, astronomers observing from the ground are taking advantage of the continual march of technological progress in electronics to build ever more sensitive detectors in the race to construct the full and detailed radiation signal from the first moments of the universe's history. Did inflation really happen? The eventual form of Figure 9.6 could be what ultimately persuades cosmologists. It reveals the rhythm of inflation. It allows us to see right back to when the universe was little more than 10^{-35} seconds old. Frequently reproduced in the science news media and in scientific and public talks about the state of play in modern cosmology, one day it will be seen as the first evidence we found about the earliest conceivable moments of our universe. It is the cosmological counterpart of a baby photo.

ETERNAL INFLATIONARY UNIVERSES

'The World is Not Enough'
    James Bond's family motto

The inflationary picture of the early history of the universe gave rise to two unusual elaborations that were entirely unsuspected. It proposes that the universe underwent a brief surge of accelerated expansion very early in its history. This enables the whole of the portion of the universe that is visible to us today to be the expanded image of a region that was once small enough to be smoothed by light rays and other processes that arise naturally in the early stages of the universe. At first, our attention was focused almost completely on how success-fully this simple proposal explained the gross properties of our visible universe: its overall uniformity, peppered with the small irregularities destined to become galaxies, its special expansion rate and the extreme similarity in the expansion rate from one direction to another. All had previously been unexplained coincidences. Now they were conse-quences of a single simple hypothesis.

However, this simple picture soon revealed some unexpected compli-cations. It is all very well to consider our visible portion of the universe as the smoothly expanded image of one tiny primordial region – but what about all the neighbouring regions? Each of them might undergo a slightly different amount of inflation and create large smooth regions of their own, but with different properties to our own. We can't see them because their light has not had time to reach us yet,[24] but one day trillions of years in the future our descendants might glimpse them, and what they will find is a region of the universe whose geography is very different to our own. We expect the whole universe to be enor-mously complicated and irregular over very large scales even though it looks very smooth and relatively simple over the scales that are visible to us today. Inflation leads us to expect that the local universe that we

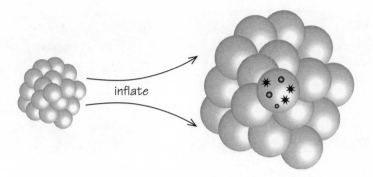

Figure 9.7 Chaotic inflation. Different patches of the early universe undergo different amounts of inflation in a random fashion. We find ourselves in one of the patches that grew large enough, and old enough, for stars to form and carbon-based life to evolve.[25]

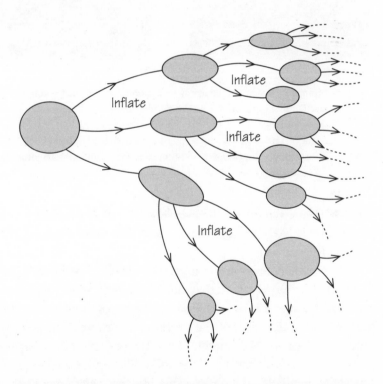

Figure 9.8 Self-reproducing eternal inflation. When each patch inflates it creates conditions for parts of it to undergo more inflation in the future. This process appears to have no end and may have had no overall beginning.

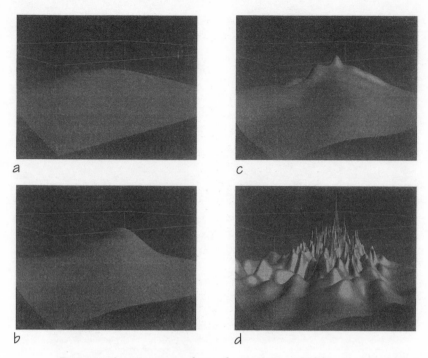

Figure 9.9 Successive snapshots of a computer model created by Andrei and Dimitri Linde to follow the development of self-reproducing inflation, from frame (a) to frame (d). The formation of hills in space signals the onset of inflation. The subsequent appearance of hills on hills, like stalagmites, shows the runaway self-reproduction of inflationary universes.

can see is not typical of the whole (and possibly infinite) universe. Geography is likely to be a much more complicated subject than we thought (Figure 9.7).

If this wasn't unnerving enough, it was then recognised by Alex Vilenkin and Andrei Linde that the inflationary universe has another awkward feature: it is *self-reproducing* (Figure 9.8). Once inflation takes hold in one small region of the universe, and causes it to accelerate its expansion, it creates the conditions needed for further inflation to occur in sub-regions of this first inflating region. The result is a self-reproducing process with inflating regions spawning further inflating regions, which in turn produce others, and so on, *ad infinitum*. And if *ad infinitum* to the future, then why not to the past as well?

Eternal self-reproducing inflation means that while our little

Figure 9.10 Sol Lewett's *Splotch 15*, 2005. This conceptual art installation is 12 feet tall and was made using acrylic on fibre glass. It is one of a series of twenty-two evolving structures in similar style begun in 2000.

inflating 'bubble' universe may have had a beginning when its expansion began, the whole 'multiverse' of bubble universes need have had no beginning and will have had no end. We inhabit one of those (perhaps rare?) bubbles in which expansion persisted for long enough to allow stars, planets and life to evolve. History is also a lot more complicated subject than we thought.

This picture of 'eternal inflation' introduces a new complexity into our understanding of the history of the universe. It is tantalising because, as with the recognition of the geographical complexity of an inflationary universe, we find ourselves faced with the likelihood that we live in a universe of huge diversity and historical complexity, most of which is totally inaccessible to us. We inhabit a single, simple patch of space and time in an elaborate cosmic quilt.

Andre Linde, one of the architects of this conception of the eternal inflationary universe, created a vivid picture of the self-reproduction process in action (Figure 9.9). In these computer simulations created by his son Dimitri, the landscape of space begins to develop low hills as different amounts of inflation occur in different places. Then spikes form on the hills, and spikes on the spikes, and so on, as inflation self-reproduces in a fractal display of complexity growth. We live in one of the rare pinnacles where inflation has been completed and the expansion calmed down. Yet this appears to be an atypical site to inhabit. Most of the space in an infinite multiverse should still be undergoing inflation. This colourful sequence of inflationary stalagmites, which Linde characterises as the 'Kandinsky universe' (although they are more reminiscent of one of the 'Splotch' sculptures of the American conceptual artist Sol Lewitt, shown in Figure 9.10), is a snapshot of the endlessly bifurcating multiverse.

SUDDENLY, THE UNIVERSE SEEMS SIMPLER AGAIN

'The spreadsheet created the "what-if" society. Instead of moving forward and progressing normally, the what-if society questions every move we make. It second-guesses everything.'
    John C. Dvorak[26]

The inflationary universe has been extraordinarily successful in supplying a simple explanation for the nature of the universe that we do see, while seeking to persuade us that there is an infinitely complicated vastness beyond our horizon that we will never be able to see. The growing observational evidence for the distinctive pattern of temperature variations in the microwave background radiation means that we take very seriously the idea that our visible portion of the universe underwent a surge of inflation in its very earliest stages. This is why we now find it to be so smooth, isotropic, close to the critical rate of expansion, free of magnetic monopoles and seeded with the distinctive irregularities needed to grow into galaxies by the process of gravitational instability that Newton first pointed out and Lifshitz confirmed would also occur in Einstein's theory of gravity.

The pattern of expansion that inflation predicts is very similar to one of universes that Lemaître studied eighty years ago. What is different is that there is a way of creating a transient cosmological constant – one that decays away quickly into radiation – which leaves an observable trace. We think that we have observed those traces today. This is a fantastic state of affairs. Something could happen when the universe is just 10^{-35} seconds old that leaves a fossil remnant which we can see today. We can check our theories about what went on at that absurdly early time by direct astronomical observations.

Yet, despite its successes in explaining the universe we do see, the inflationary-universe theory spooks us by making it seem inevitable that we are part of an infinitely complicated universe of expanding 'bubbles', each probably very different in structure and possibly under different natural legislatures. Suddenly, there is no longer only one universe. There are many.

THE MANY UNIVERSES RUN OUT OF CONTROL

'As I have discovered, there are no disasters, only opportunities.
And, indeed, opportunities for fresh disasters.'
    Boris Johnson[27]

The unusual profligacy of an inflationary universe in taking tiny fluctuations in space and rapidly expanding them to the scale of entire universes forces us to start thinking about universes in the plural. We can see the expanded image of one tiny primordial fluctuation. Yet, all around us, beyond our horizon, are the results of innumerable bouts of self-reproducing inflation. What we see in our part of the universe depends on two things: the fine details of the inflation it once experienced and the laws of physics that govern it. Until fairly recently, most physicists expected that there would be only one possible 'Theory of Everything'. Finding it was like doing a big jigsaw puzzle. There is only one solution and you'll know it when you find it.

Gradually, confidence in this simple expectation has been eroded. The best candidates for Theories of Everything have an entirely different character. They have an overall symmetry that shapes the laws of Nature but they possess an unexpected degree of flexibility.

There can exist different self-consistent laws of physics, each with different numbers of forces, constants of Nature, numbers of dimensions of space, and much else besides. Many of these traditionally fixed aspects of the universe appear to be just outcomes that fall out at random. The set of laws and forces that we see in Nature is just one of the many choices of bye-laws. All are self-consistent and complete in their own way.

At first, it appeared that there might be only a few options to choose from a suite of possible Theories of Everything, called 'string theories', which were discovered in the early 1980s. But then it was discovered that these string theories were not the final story. They were not true Theories of Everything. Instead, they were merely different limiting views of a deeper, and still as yet unfound, Theory of Everything that became known as 'M theory'. We know that this theory exists but we know little about its deep structure, only what it looks like when we describe particular limiting cases, such as when energies and temperatures are low or when gravity is very weak. Remarkably, it predicts that in a situation like this gravity will be described by Einstein's theory of general relativity. From what we know of this theory, it is extraordinarily complex and it offers a very large number of self-consistent worlds; more than 10^{500} is the favourite estimate. This number is amazingly large. A billion is only 10^9 and if you started counting at the rate of one per second on the day you were born you probably wouldn't reach a billion in your lifetime.

This 10^{500} number gives all the possible states defining the laws and constants of physics that our patch of universe could find itself in after it cools down enough for inflation to begin. This realm of possibilities is known as the 'landscape'.[28] These possible worlds don't just differ slightly in density and temperature, or how they are expanding, like the different universes we have been looking at so far in this book. They are more radically different; they have different numbers of forces and constants of Nature, even different dimensions of space and time. Many may not display forces like electromagnetism and will contain no atoms or life.

The study of this vast array of possibilities is still in its infancy and involves the discovery and classification of complex mathematical structures called Calabi–Yau manifolds, after the two mathematicians, Eugenio Calabi and Shing-Tung Yau, who discovered them long before

Figure 9.11 A Calabi-Yau space.

physicists became interested in them.[29] At first, this study appears daunting. Not many of these Calabi–Yau spaces are known and understood; a picture of one is shown in Figure 9.11. They can be classified in terms of some of the defining quantities which measure their size and complexity.

Researchers have made good use of fast computers to search and classify the possibilities that exist but have only been able to explore the sparser regions of the landscape where things are not too complicated. One of the hopes for these studies is that there may only be small numbers of 'interesting' worlds among the unimaginably vast number defined by all the Calabi–Yau spaces. Only a few may allow the sub-patterns needed for physics to accommodate particles like electrons and quarks, and combinations of them that lead to atoms and molecules.

In the meantime, we are faced with an extraordinary cosmic lottery in which it seems that each of the different inflating patches of an eternally inflating universe will find itself dropped at random into one of the 10^{500} available states in the string landscape of possibilities. After inflation, the state that was chosen by a haphazard sequence of events will determine the character of the large inflated region that results.

The number of different types of physics available in the landscape sounds staggeringly large and the exploration of parts of the huge (but finite) catalogue of Calabi–Yau spaces appears increasingly challenging. It will keep mathematicians busy for a long time. However, a deep problem about the systematic exploration of all the possible states of the landscape has been pointed out by Frederik Denef and Michael Douglas, at Rutgers University.[30] One might have thought that the strategy to be adopted by physicists would be to determine the places in the landscape which lead to universes like the one we observe. Unfortunately, this idealised programme falls foul of what we know about the complexity of computational problems. Any checking process for the lowest energy states of the string-theory landscape is computationally so long-winded ('NP-hard' is the technical term) that identifying and checking the properties of all the candidate states of the theory is beyond the reach of any conceivable computer (even a quantum computer).

These 'hard' problems are defined to be ones where the amount of computation required grows exponentially with the input information. We are used to handling 'easy' problems where the amount of computation grows in direct proportion to the size of the input (or to some mathematical power of it). Some simple searches have been done[31] which scanned 100 million models of the theory and used up 400,000 hours of computer time (many machines can be used simultaneously). Yet the complexity and computation time grow so fast with the size of this 'hard' problem that if one just doubles the number of models to be explored, the computer time needed would grow to more than 100 million years.

This large number of possible outcomes for the inflationary universe is alarmingly large. We are not going to be able to use any sequence of observations to eliminate the unsuitable candidates one by one. This sounds worrying, but things are actually a lot worse. So far, we have been counting the number of different types of physics that string theory permits, but we haven't begun to count how many inflationary 'universes' the eternal inflationary self-reproduction process will produce.

Suppose that we take a little patch of space that has randomly chosen one of the superstring landscape vacuum states to reside in. This picks out the physics that will govern it. Its size will be at most equal to the speed of light times the age of the universe when it begins to inflate. In order to inflate enough to encompass our visible universe today, and so explain its isotropy and smoothness and lack of magnetic monopoles, it will need to grow by a factor of at least $N = e^{60}$ when it inflates. The number of other 'universes' with different geometrical properties which will subsequently be produced as by-products of this process will be at least as large as

$$e^{e^{3N}} = 10^{10^{77}}$$

This number is so large that our minds would be unable to store all the information needed to list all these possible worlds. It would take all the atoms in our visible universe just to have enough ink to write it out in full as 10 followed by 10^{77} zeroes. It is vastly bigger than the number of members of the string vacuum landscape. In fact, we can make the number even bigger if we don't just count the regions that grow from our patch but include those that emerge from a larger part of the universe. The entire universe is an ever-growing, expanding fractal branching process, budding an amazing number of new regions all the time.

We don't need to produce any more big numbers to get the message across. The number of potential universes and the number of actual universes that are as big as the region we call 'our visible universe' today is mind-bogglingly large.

10 Post-Modern Universes

'When you come to a fork in the road, take it.'
    Yogi Berra

RANDOM UNIVERSES

The Cosmos is about the smallest hole that a man can hide his
head in.
    G. K. Chesterton

Throughout the twentieth century cosmologists delved into many
possible universes and used astronomical observations to pick the one
that best fitted the facts. Now they think about lots of universes all
at once – a 'multiverse' of possibilities – each occupying a region
within the entire universe bigger than the whole of our visible universe.
What is new is that these possible universes may all actually exist
somewhere in real space right now. They are not merely the 'possible
universes' of the philosophers, the might-have-been universes of the
virtual historians, or the what-if universes of the Olympic silver
medallist.

Look back at the eternal inflationary universe. The continual budding
of new universes in a self-reproducing process is expected to continue
for ever and it may have had no beginning. If we could take a God's-eye
view we would find that most of the multiverse should still be undergoing inflation. In the minority will be regions like our own where
that first burst of inflation has ended and normal decelerated expansion resumed. Different regions will expand in different ways, with
different densities, temperatures and shapes. In some, inflation may

have been too brief to drive the expansion close to the critical divide; in others, the irregularities generated from stretching the quantum fluctuations might have been much weaker or much stronger than those we see in our patch. And each region could have its own physics as well.

How do we deal with all these possibilities? There are an infinite number of possible universes. The number is too large to be explored systematically by any computer. The choice of vacuum or the length of each bout of inflation cannot be predicted by solving some simple equation, or even a complicated one: it is *random*.

Randomness means different things to different people. To some it means disorder. To others its means uncertainty, or the source of an inability to determine something precisely. To others still, it means total unpredictability. In the early universe all the events that are categorised as random, and can be predicted only in terms of the probability that they will occur, owe that indefiniteness to their quantum origins. Their randomness derives from the uncertainty inherent in the quantum nature of matter and energy. This uncertainty cannot be reduced by gathering more or better information. It is an uncertainty intrinsic to the concepts, like space and time and motion, that we use to describe the world. Two identical causes will not have the same quantum effects.

When random events occur during the very early history of the universe they have effects that reach into the far future. The little variations in density that seeded the formation of great clusters of galaxies owe their beginnings to quantum uncertainty and randomness at the time when inflation occurred.

This quantum graininess means that when we seek to predict the course of events in an eternally inflating universe we need the language of probability. Ideally, if you came along and specified a type of universe in general terms – say, one rather like our visible universe – we would like to know how likely it is that such a region would be generated in the self-replication process of eternal inflation. Unfortunately, so far this has proved too hard a problem for cosmologists to solve. The definition of what you mean by 'probability' or 'likelihood' in this situation has proved too difficult to pin down.[1] However, the situation is far from hopeless and some different possibilities have been proposed and are being intensively explored. Each has so far proved to have a weakness. We are perhaps just one good idea away from solving this problem.[2]

PROBABLE UNIVERSES

'I am searching for abstract ways of expressing reality, abstract forms that will enlighten my own mystery.'
Eric Cantona

What does it mean to say our visible universe is 'improbable' or 'probable'? Suppose that we had found that the probability of there being a universe with some attributes (containing atoms or stars, for example) had the shape of Figure 10.1. The graph we have drawn has a typical form for a spread of probabilities for the value of some 'constant' of Nature. There is a peak, or most probable outcome, and the probability becomes smaller and smaller as we move away from this outcome in either direction. This picture might be a prediction of our multiverse theory and we want to see how to use it to test the theory. Would we expect our visible universe to display the most probable value predicted by the picture? More crucially, would we regard the theory as false if we found our visible universe displayed a value for this constant that our picture showed to be a very improbable outcome in this theory?

You would be mistaken if you answered 'yes' to either of these questions. In order to interpret the picture we need to know something about 'life' in the multiverse. It might be that life of any sort, perhaps defined rather minimally as 'atomic complexity', is only possible if the 'constant' takes a particular range of values. If the most probable outcome from eternal inflation falls outside this range then no life is possible in the most likely universe (perhaps because the temperature never falls below a million degrees), we should not be surprised we don't observe a 'most probable' universe. Moreover, if the range of values for the constant that does allow life to exist in a universe is narrow, and very improbable, we should not exclude our theory on the basis of this picture. *What matters is not the prediction for the probability of a universe having a particular property but the conditional probability that it has such a feature given that it is also possible to have*

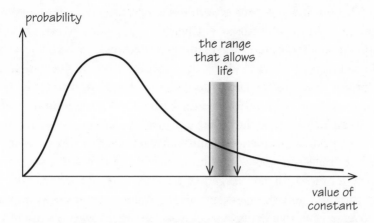

Figure 10.1 A possible form for the probability that a constant of Nature takes different possible numerical values in a universe. The most probable value may fall outside the range that permits life to exist and so we could not observe the most probable universe.

observers ('life') in the universe. Universes that don't produce the possibility of 'observers' – and they do not need to be like ourselves – don't really count when it comes to comparing the theory with the evidence.

This is rather sobering. We are not used to the existence of cosmologists being a significant factor in the evaluation of cosmological theories.

ANTHROPIC UNIVERSES

We would be better mirrors of the Universe if we were less concerned about our own image.
   Maurice Druon[3]

Cosmologists have long appreciated that there is a surprisingly close link between the large-scale properties of the universe and the existence of life within it. At first sight this might appear odd. The universe is big. There are countless stars and galaxies extending billions of light years in every direction. How could any of that have a connection with us, here and now, orbiting an average star in an unremarkable galaxy?

The unexpected connection arises because the expansion of the universe links time and space. Chemical elements like carbon, oxygen and silicon that produce the complex structures on which 'life' is based do not appear ready-made in the universe when it starts expanding. Nor are they made during the primordial nucleosynthesis of deuterium, helium and lithium during the first three minutes of the universe's history. Rather, they are made in the stars over billions of years through a sequence of nuclear reactions that first combine two helium nuclei to make beryllium, then add another to make carbon, then another to make oxygen, and so on. These processes take place as stars die and the elements they produce are spread around the universe when the dying stars explode as supernovae. Eventually, they find their way into dust and rubble that condenses into planets, and then find their way into molecules and people.

The production of carbon and steady hydrogen-burning stars like our Sun that sustain life-supporting environments needs billions of years of stellar alchemy to take place. This is why we should not be surprised to find that our universe is so old. It takes lots of time to produce the chemical building blocks needed for any type of complexity. And because the universe is expanding, if it is old, it must also be big – billions of light years in extent. If the universe were just the size of the Milky Way galaxy, with its 100 billion stars offering potential homes for planetary systems, it might seem room enough for lots of life. But such an economy-sized universe would be little more than a month old. No time for stars to evolve or for the building blocks of biochemical complexity to be formed.

If we look at other aspects of the visible universe we find other possible features that need to fall in particular ranges if life is to be a possibility. We have seen that universes can expand faster or slower than the critical rate. If they had started expanding much faster when the universe was younger, no galaxies and stars could have formed because material would have been dragged apart too fast for the gravitational instability process of Newton, Jeans and Lifshitz to capture it. By contrast, if the expansion had been a lot slower than the critical rate, material would have coalesced into dense lumps very early in the history of the universe and we would have a universe of black holes instead of hydrogen-burning stars. Again, we see that we should not be surprised to find that our universe

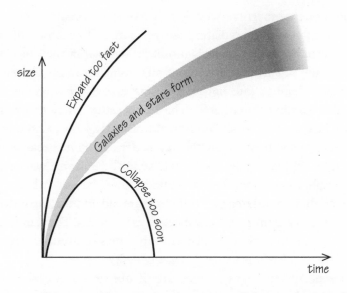

size

Expand too fast

Galaxies and stars form

Collapse too soon

time

Figure 10.2 Universes that do not expand close to the critical divide will collapse to a big crunch before stars can form, or expand too rapidly for material to overcome the expansion of the universe and coalesce to form galaxies and stars.

expands at a rate that is close to critical. We could not have existed if it hadn't.

The inflationary-universe proposal provides a simple explanation for why a post-inflationary universe will be found to expand very close to the critical rate for a very long time (Figure 10.2).[4] This way of thinking looks strange at first. It is usually called the 'Anthropic Principle'. It is not a theory that can be proved true or false – it is true – but just a methodological principle which will stop you drawing wrong conclusions from the evidence. It recognises that there are properties of the universe which are necessary for the evolution and persistence of life within it and so we could not observe them to be otherwise. Many can be traced back to the need for enough time to make biochemical elements, which we have just seen means the observable universe must also be big. It also means that it must be very cool and dark. Billions of years of expansion mean that the temperature of radiation in space will steadily fall, halving every time expansion doubles the scale of the universe. This expansion continually lowers the density of matter in the universe so that today there is only about one atom in every cubic metre of space on average. The density is so small and the radiation temperature so low after billions of years of expansion that there is too little material in the universe to illuminate the sky around us. Even if all the matter in the universe were suddenly changed into light using Einstein's $E = mc^2$ recipe we would notice nothing remarkable. The temperature of radiation in space would experience a negligible rise from 3 degrees to about 15 degrees above absolute zero. So the sky is dark at night because of the great age of the universe. The night sky in the universe was once bright, as bright as the surface of the Sun all over the sky, when the universe was about a quarter of a million years old and the expansion was about a thousand times less advanced than it is today. The temperature was too high for any stars, planets, molecules or atoms to exist. No observers could witness that bright night sky.

This new perspective on the structure of our universe, and other universes we could imagine, unifies our understanding of the connections that exist between life and the cosmos. It can also stop us drawing wrong conclusions from the astronomical evidence. In chapter 4 we saw how Paul Dirac had sought to explain some numerical coincidences between different constants of Nature by equating two combinations of them which were roughly equal to 10^{39}. He thought the numbers were so unusually large that their similar magnitudes must mean that they were linked by some undiscovered equation, or law of Nature. As we saw earlier, the novelty of his suggestion was that one of the large numbers was built from quantities that were regarded as true constants of Nature; the other involved the present age of the universe, which Dirac also included as one of his 'constants'. But the age of the universe changes as time passes and so by equating his two 10^{39} numbers he forced one of the other traditional constants of nature to change with time. He picked Newton's gravitational constant, G, and required it to fall in value as the universe aged, inversely with time. This was a radical suggestion. But it was unjustified. Robert Dicke pointed out that the coincidence between the large numbers is just another way of stating that the time during which we are observing the universe is roughly the time it takes for a star to settle down and burn hydrogen in a steady fashion. Since we would not expect to be on the astronomical scene before the stars form and observers are unlikely to be around after they all exhaust their fuel,

Dirac's coincidence wasn't unlikely at all. By ignoring the role of the observer in biasing our observations he drew the wrong conclusion. In a big-bang universe there are preferred periods of cosmic history, when stars can form and remain in stable existence, when atoms can exist, and so forth. The steady-state theory had tried to force the universe into a straitjacket where there are no preferred times in cosmic history, but the astronomical evidence told strongly against it.

Dirac was publicly unmoved by Dicke's argument and claimed that although he was happy to believe we would not be observing the universe before the stars formed, he could see no reason why there should not be observers after the stars have exhausted their fuel:

On Dicke's assumption habitable planets could exist only for a limited period of time. With my assumption [i.e. varying G] they could exist indefinitely in the future and life need never end. There is no decisive argument for deciding between these assumptions. I prefer the one that allows the possibility of endless life.[5]

Remarkably, Dirac's attraction to the idea of unending life in the universe is one that was expressed in remarkably similar words in January 1933 when he decided to set down his philosophy of life in just three notebook pages. Writing not long after the suicide of his brother Felix, he rejected conventional forms of religious belief but replaced them with another type of faith to give meaning and purpose to human life. His private notebook sheds some background light on his subsequent cosmological perspective:

[My] article of faith is that the human race will continue to live for ever and will develop and progress *without limit*. This is an assumption that I must make for my peace of mind. Living is worthwhile if one can contribute in some small way to this endless chain of progress.[6]

Despite appearing publicly unimpressed by considerations of the role of observers in conditioning cosmological observations, Dirac said rather different things in his private correspondence with Gamow. As we have seen, Dirac's simple varying-G theory was soon ruled out by consideration of the consequences for the evolution of the Sun and the temperature of the Earth if G had been much bigger in the past. One rather unlikely scenario put forward in its defence by Dirac was to imagine that the solar system had occasionally passed through huge clouds of dust as it orbited the Milky Way galaxy. This led to the accretion of extra material by the Sun which would increase its mass and offset the effects of falling G on its gravitational pull on the Earth. Gamow thought this was hopelessly improbable and 'inelegant' – a charge meant to resonate with Dirac's legendary desire for physics to use and express beautiful mathematics. Why should the extra mass accreted be just the right amount to cancel out a cosmological varying G? Yet Dirac defended his idea by appeal to an anthropic argument, in his surprising letter to Gamow on 20 November 1967:

I do not see your objection to the accretion hypothesis. We may assume that the sun has passed through some dense clouds, sufficiently dense for it to pick up enough matter to keep the earth at a habitable temperature for 10^9 years. You may say that it is improbable that the density should be just right for this purpose. I agree. *It is improbable.* But this kind of improbability does not matter. If we consider all the stars that have planets, only a very small fraction of them will have passed through clouds of the right density to maintain their planets at an equable temperature long enough for advanced life to develop. There will not be so many planets with men on them as we previously thought. However, provided there is one, it is sufficient to fit the facts. So there is no objection to assuming our sun has had a very unusual and improbable history.

Curiously, Dirac never used this type of reasoning as a way of understanding his Large Number coincidences.

In retrospect, we can see that this type of 'anthropic' reasoning would also have made the steady-state cosmology look rather questionable in the 1950s. In the big-bang picture of the universe, the age of the universe is roughly equal to the reciprocal of the expansion rate of the universe, or 'Hubble's constant', as astronomers call its presently measured value.

In the steady-state theory, there is no finite age for the universe (it is infinite) and its expansion rate is a completely independent property of the universe needing a separate explanation. The observed fact that the ages of stable stars like the Sun are very close to, but obviously a bit less than, the age of the universe is entirely natural in the big-bang model. Galaxies form, then stars, then planets, and then astronomers, in a linked historical sequence.[7] Hence, the fact that the expansion rate of the universe today is roughly the inverse of the age of stars is not surprising in a big-bang universe.[8] In the steady-state theory it is a complete coincidence.

Astronomers steadily became familiar with the sensitivity of various constants of physics to the existence of life in the universe. Just as little changes in the expansion rate of the universe had major consequences for life, so changes in the strengths of the forces of nature or the masses of elementary particles could stop stars or atoms existing and change the course of cosmic history.[9] Such considerations of the sensitivity (or insensitivity) of the constants of physics and the structure of the universe for life became known as 'anthropic' arguments.[10] They have sometimes led to claims that the universe we observe is 'fine-tuned' in certain respects that are conducive to the evolution of life. If values of some constants were changed a little then the window of cosmic opportunity for the formation of atoms or stars and the evolution of biochemical complexity would be closed.

There are many difficulties in determining the meaning of these observations. How broad should our definition of 'life' be? What do we mean by 'little' when it comes to changes to constants of Nature? Are all these constants really independent of each other or is that just an artefact of our lack of knowledge of a fully unified theory of physics?

Until the late 1980s this perspective on reality seemed rather eccentric. Most cosmologists thought that there is one universe with the properties it has; there is nothing more to say at a scientific level. You could go a little further by imagining that there exist many (even all) possible universes in some metaphysical sense and then place ours within this gallery in the one neighbourhood where life is possible.[11] You can try to interpret the situation in the light of philosophical perspectives or religious views about whether the universe might be well suited for life. But if the universe was a one-off and if it hadn't been suitable for life we wouldn't be here to talk about it.

A key presumption of the only-one-universe perspective was that all the constants of Nature and every one of the defining properties of the universe are uniquely and completely defined. There is no flexibility for another universe that has slightly different laws and constants

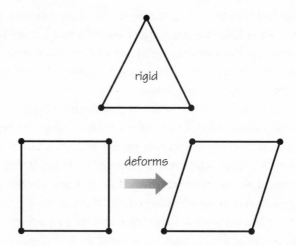

Figure 10.3 A triangle made of hinged struts is rigid: it cannot be deformed smoothly into another shape without buckling. A hinged square is not rigid: it can be smoothly deformed into a parallelogram.

of Nature. Imagine making a triangle and a square out of Meccano struts with bolts at the corners. They are very different. The square is not rigid: it can be shifted slightly by pulling the top to the right and the bottom to the left to change it smoothly into a parallelogram (Figure 10.3). But you can't do this with the triangle. It is rigid.[12]

Are the laws and constants of physics like the rigid triangle, with no similar set nearby? Or are they like the flexible square with an unlimited number of alternatives, some very similar, and others rather different? Even in the case of the square there are still constraints imposed by the joints and lengths of the sides. Likewise, even if there are different sets of constants or laws allowed, that doesn't mean *anything* is allowed. There can still be an overarching constraint that lies deeper than the measurable values of the constants of Nature. Pioneering physicists like Einstein who sought a grand Theory of Everything (his 'unified field theory', or Eddington's 'fundamental theory') believed strongly in the uniqueness of the ultimate description of the world that physicists could achieve. Indeed, this was the rationale for trying to find it by appeal to mathematical symmetries and pure thought. It was far beyond the reach of experiments at that time. You had to believe that there was a beautiful mathematical structure that would leap out of the page and say: 'This is the only way it can be!'

If you had asked Einstein about other universes, or some type of multiverse, where the constants of Nature took different values to those we observe, he would not have been very interested. He once wrote to an old friend and life-long correspondent, Ilse Rosenthal-Schneider, that:

In a reasonable theory there are no dimensionless numbers whose values are only empirically determinable. Of course, I cannot prove this. But I cannot imagine a unified and reasonable theory which explicitly contains a number the whim of the Creator might just as well have chosen differently, whereby a qualitatively different lawfulness of the world would have resulted . . . Dimensionless constants in the laws of nature, which from the purely logical point of view can just as well have different values, should not exist. To me, with my 'trust in God' this appears to be evident, but there will be few who are of the same opinion.[13]

POSSIBLE UNIVERSES

It was a time when everything was supposed to change, but everything stayed the same, only in a different way.
    Henning Mankell[14]

Since 1990 there has been a steady erosion of the old belief that the universe and its defining constants and laws are like a rigid triangle. Many of the properties of the universe that once seemed hard-wired were now seen as the outcomes of symmetry-breaking processes respecting deeper principles. If you cool down a bar of iron from a very high temperature it can become a magnet if the temperature falls below about 770°C. There is an equal chance that either of its two ends will become the north pole and the other the south pole. Which one will be north is not something you can predict in advance of cooling; when the temperature falls to 770°C the atomic symmetry in the iron breaks down and a direction of magnetisation will be randomly chosen.

This type of random 'symmetry-breaking' process can determine many vital aspects of the universe – the matter–antimatter imbalance, the density of atomic matter, and even the strengths and orientations of cosmic magnetic fields. In these circumstances the anthropic perspective becomes very different: if these symmetry breakings can occur in different ways in different regions of the universe, the different outcomes may simply describe a new variability in the universe's properties.

The unusual behaviour of the eternal and chaotic inflationary universes shows why it is vital to take anthropic selection into account. There are so many regions which are larger than our visible universe that we need to work out the likelihood of generating a universe with a particular spectrum of astronomical properties. If we add in the complexity contributed by symmetry breaking we need to recognise that the different 'universes' spawned by eternal inflation could also display very different physics. Many of the constants of Nature could fall out differently in different regions; string theories suggest that the number of large dimensions of space (and even of time) could vary within the multiverse; and there could be different collections of fundamental forces, each of these differences reflecting a different choice of vacuum state in the string landscape.

This is rather daunting for the traditional scientific method. Our anthropic considerations have taught us that in making sense of the likelihoods of different universes arising in this multiverse of actualities we need to confine attention to those that can give rise to complexity, life and conscious observers at some stage in their evolution. That subset includes our visible universe but we don't want to specify life too conservatively, so that there is no room for forms of consciousness that are totally unlike us. At present, it seems far too hard to specify the most general conditions needed for a universe to support observers. We don't even know the necessary conditions needed to define 'life'. We know only some weak sufficient conditions that are informed very closely by what we know about ourselves.

This makes it very difficult to predict what the most likely type of life-supporting universe in the multiverse will be. In order to do so we need to know every requirement on the physics of matter and the structure of the universe that impinges critically on the existence of biochemical complexity. We know some of these critical factors but we can be sure that the true restrictions are greater than we currently think. This is because we have not yet found the Theory of Everything which unites the four forces of Nature (electromagnetism, weak, strong and gravitational). When that theory is found it can only increase the number of interconnections and dependencies that exist between the constants of Nature that characterise those four forces. At present we can imagine the effect of changing rates of radioactive decay without worrying if there are consequences for gravity or atomic structure. A fully unified theory would reveal all those interconnections and ensure that a small change in one part of physics would have extra consequences elsewhere.

If we solved the problem of how to calculate probabilities and conditional probabilities – which I am sure we will in the not too distant future – we are faced with the vast complexity of tracking down all the dependencies of 'observers' on physics and having a full theory that expresses those dependencies. This approach might provide us with several different families of universes which meet our necessary requirements for life. Next, we can ask which are the most likely. But what if the most probable family doesn't contain a universe that looks like the one we see? Would we conclude that some part of the theory (or all of it) was false, or would we conclude that we don't live in the most probable universe? And if we found several families, all roughly as likely as each other, how would we make a choice as to which of the universes was preferred?

These questions, and the lack of clear answers to them, should be seen as reasons for keeping these anthropic considerations strongly in mind. Whatever the answers to our questions about the likelihood of universes with different properties arising in the multiverse, they will inevitably involve incorporating our ideas about observers. We will have to come to terms with the fact that in the last analysis we (and other entities that process and gather information) are part of the problem we are trying to solve.

HOME-MADE UNIVERSES

> The point of philosophy is to start with something so simple as
> to seem not worth stating, and to end with something so para-
> doxical that no one will believe it.
> Bertrand Russell[15]

The picture of an eternally self-reproducing universe exploring the suite of all possible vacuum states of the universe, with their different constants of physics, dimensions of space and characteristic forces, is the nearest we have to an exploration of all possible worlds. Philosophers have often speculated about the metaphysical realm of all possible worlds and there have been heated arguments about whether we can be said to inhabit the 'best of all possible worlds' in any meaningful sense.[16] While the inflationary universe doesn't create 'all' possible worlds in the metaphysical sense of creating all imaginable variations of all aspects of a universe, it is well able to explore all the self-consistent editions of physics that string theory allows. This appears to be the huge, but finite, number – 10^{500} – we met earlier.

The reproduction process is described by the laws of physics. It is not a mysterious 'creation out of nothing' like the creation of the universe traditionally enshrined in medieval Christian theology. At first, it might sound impossible to 'create' something, even in this more limited sense. But no conservation laws of physics are violated. Imagine that a quantum fluctuation produces a particle and an anti-particle spontaneously. If each has a mass, M, this requires a total amount of positive energy equal to $2Mc^2$. But if those particles exert an attractive force on each other – for example, gravitational or (if they have opposite electric charges) electromagnetic – this contributes a 'potential' energy that is *negative* because it is available to create energy of motion if released, just as when you drop a stone and it falls to the ground under the gravitational pull of the Earth.

Remarkably, the sum of the positive energy $2Mc^2$ and the negative potential energy of attraction can total zero and the 'creation' of the two particles out of the quantum vacuum need cost no energy at all (Figure 10.4).

This realisation, together with the picture of the natural self-reproduction of 'universes' during eternal inflation, led cosmologists to ask whether it might be possible to stimulate this production of a universe artificially.[17] Could we 'create' a universe in the laboratory by stimulating one of the fluctuations that produce the same effect in the eternal inflation process? Do the laws of physics allow it, even in principle?

Several attempts were made to prove that this was possible or impossible, none definitive, but there seemed to be dramatic unwanted by-products – like infinite densities. At the same time, there were others, like the late Ted Harrison at the University of Arizona, who

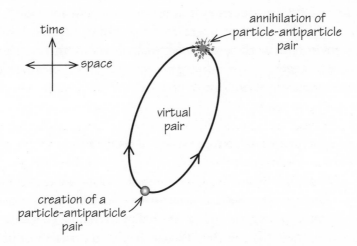

time

space

annihilation of
particle-antiparticle
pair

virtual
pair

creation of a
particle-antiparticle
pair

Figure 10.4 A virtual particle–antiparticle pair appears and annihilates.

saw how this opened up the bewildering prospect of advanced civilisations manipulating the universe.[18]

Imagine very advanced civilisations in the universe that have developed a much fuller understanding of the possibilities for creating special fluctuations in their own part of the universe, which then inflate rapidly to produce new baby universes. These super-cosmologists will also have a good understanding of all those quirks and apparent coincidences in the values of the constants of Nature and the laws that govern them which make life a possibility. If – and it is a big if, given the number of possibilities and the computationally 'hard' nature of selecting them – they are able to control the symmetry-breaking process that determines the vacuum state that is chosen as the temperature falls, they would be able to force-breed new universes in which the constants and laws of physics were more conducive to life than in their universe. Fast-forward through several generations of this process. More advanced civilisations would arise (more easily, one assumes, because life-supporting 'coincidences' have been fine-tuned further for their universe by their predecessors) with ever-increasing capability to fine-tune the universes they make. Perhaps, Harrison speculates, the fact that our universe displays many apparent fine tunings that are advantageous for the evolution and persistence of life indicates that this sort of forced breeding[19] of universes has gone on in the past and is responsible for the fine tunings we witness today.

This is a fantastic idea but it brings out a remarkable point. Once universes evolve forms of intelligence that can manipulate the large-scale environment, then cosmology can become an unpredictable science like economics or sociology because human choices are unpredictable in principle, not just in practice.

NATURALLY SELECTED UNIVERSES

There is a theory which states that if ever for any reason anyone discovers what exactly the universe is for and why it is here it will instantly disappear and be replaced by something even more bizarre and inexplicable. There is another that states that this has already happened.

Douglas Adams[20]

Another attempt to address the puzzle of the special values of constants of Nature was made by the American physicist Lee Smolin. This differed from Harrison's fantastic scenario by not seeking to incorporate the consequences of intelligent observers steering home-made universes along life-enhancing histories. Instead, Smolin looked at a variant of the old oscillating-universe idea of Tolman's wherein a contracting closed universe bounces back into an expanding universe.[21] Smolin wanted to apply this notion to the gravitational contraction that occurs inside black holes. After all, if you are inside a big black hole it is the same as being inside a closed finite universe for all practical purposes. If you travel out towards the black-hole boundary from its centre you can never pass out through it (Figure 10.5). You always find yourself drawn back into the singularity at the centre from whence you came. It is just like expanding away from the singularity, reaching the maximum size and then contracting back to a singularity in a closed universe.

The first speculation was that each time a black hole forms it will lead to the spawning of new universes that 'bounce' out from the central singularity when material collapses into it. The second was to take an old idea of John Wheeler's, first suggested in 1957, that in any bounce of an oscillating universe the values of the constants of Nature might be changed – 'reprocessed', as Wheeler suggested.[22] Smolin

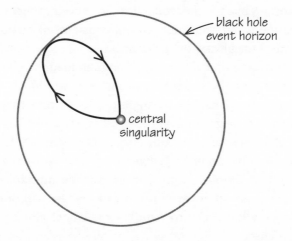

Figure 10.5 A particle launched from the central singularity of a black hole achieves its maximum distance away when it reaches the horizon, but then it inevitably falls back into the singularity. Its history is similar to that of a particle in a closed universe expanding from a big bang to a maximum size before collapsing back into a big crunch.

assumed that this bounce would happen each time a black hole formed and its trapped contents collapsed inexorably into its central singularity, with the values of the constants in the new expanding deal just slightly shifted from the old ones. The result would be a self-reproduction process in which the constants shifted slightly each time reproduction occurred.[23].

What happens if we let this process run for many generations? Black holes form in a universe and then each of those black holes leads to the formation of more universes, each with slightly shifted constants. In the long run, it was claimed, we ought to find ourselves in a universe in which the constants take values which maximise the production of black holes because that is the most likely type of universe to be in. It is not unlike biological evolution. If there are inheritable traits that lead to the greatest number of offspring then creatures with those traits will become commoner than those that don't possess them. Here, it is the gravitational trait of producing black holes that gives a universe greater fecundity.

You might be able to predict what you should observe if you lived in a universe that has resulted from many generations of this gravitational natural selection process. Since the evolution selects for maximising black hole production we would expect to be at or near a situation where the values of our constants of Nature are at a local maximum for black hole production.

This means that if a small change is made in the values of our constants then the black hole production would always be reduced. This sounds appealing. It suggested to Smolin that we carry out some thought-experiments to see what happens if we imagine small changes in the values of our constants of physics. Can we rule this theory out by finding changes that lead to increased black-hole production? Smolin's aim was to provide an explanation for the unusual values of many constants of Nature without any appeal to anthropic selection or to a multiverse where almost all possibilities were realised.

Unfortunately, this simple idea doesn't work. Even aside from the question of whether there are small changes that increase black-hole production, there are two severe problems. Although one of the aims of the theory was to free the cosmology from having to worry about anthropic influences on predictions, one can see immediately that this theory is actually a prime illustration of how important it is. We would only expect our universe to be one that maximises the production of black holes if such a universe can support observers. If universes that maximise the abundance of black holes contain only radiation and very small black holes, then no sentient beings will observe them. In order to make a testable prediction from this theory you need to know the subset of values for all the constants of Nature which allow observers to exist – but these values could be far from the most probable. Then, most likely, you could say it predicted a collection of values for these constants that maximised black-hole production within the life-supporting set of possibilities. The situation is an example of the subtlety we highlighted in Figure 10.1.

The other difficulty with this picture is more serious. It assumes that the black-hole production rate has local maxima as you change the values of constants of Nature. This is a very strong assumption and it is not true for one of the known constants. If we make little changes in G, Newton's constant that controls the strength of gravity, we find that the local geography is just a one-way slope. Instead of making more black holes, the second law of thermodynamics equally favours making more black holes or choosing a larger value of G each time the constants are shifted and there will be a steady slide towards universes with stronger and stronger gravity.[24]

FAKE UNIVERSES

> If you might be living in a simulation then all else equal you should care less about others, live more for today, make your world look more likely to become rich, expect to and try more to participate in pivotal events, be more entertaining and praise-worthy, and keep the famous people around you happier and more interested in you.
>
> Robin Hanson[25]

Once you take seriously that all (or almost all) possible universes can (or do) exist then a slippery slope opens up before you. As we have seen, technical civilisations only a little more advanced than ourselves might have some capability to simulate universes in which self-con-scious entities can emerge and communicate with one another.[26] They would have computer power that differed from ours by a vast factor. Instead of merely simulating their weather or the formation of galaxies, like we do, they would be able to go further and watch the appearance of stars and planetary systems, followed by the develop-ment of planetary geology and topography on their simulated worlds. Next, having coupled the rules of biochemistry with their astro-nomical simulations, they would be able to watch the evolution of life and consciousness (all speeded up to occur on whatever timescale was convenient for them). Just as we watch the life cycles of fruit flies, they would be able to follow the evolution of life, and watch civilisations grow and communicate with each other. They could even watch their simulated beings argue about whether there existed a Great Programmer in the Sky who created their universe and could intervene at will in defiance of the laws of Nature they habitually observed.

Once this capability to simulate universes is achieved, fake universes will proliferate and will soon greatly outnumber the real ones. And so, Nick Bostrom has argued,[27] any randomly chosen thinking being is more likely to be in a simulated reality than a real one.[28]

Motivated by this alarming conclusion there have even been suggestions as how best to conduct ourselves if we have a high probability
of being a simulated creature in a virtual reality. Robin Hanson
suggests, in the epigraph introducing this section, that you should act
so as to increase the chances of continuing to exist in the simulation,
or being resimulated in the future, although the advice sounds very
much like the personal advancement strategy adopted by many people
regardless of the type of universe they might believe they are in. In
response, Paul Davies has argued that this high probability of living
in a simulated reality is a *reductio ad absurdum* for the whole idea that
multiverses of all possibilities exist: it would undermine our hopes of
acquiring any sure knowledge about the universe.[29]

The multiverse scenario that we have introduced in connection
with the inflationary universe has been used explicitly by some cosmologists as a way to counter any argument that the universe was specially
designed for life by a Grand Designer.[30] No matter how narrow the
window of conditions needed for life to be possible, if all possibilities
are available somewhere in infinite space there will always be a habitable universe in the multiverse. Again, the key issue that is usually
ignored in this discussion is: 'What is the probability measure for that
habitable zone in the space of all possibilities?' No one knows. Others
saw the multiverse simply as a convenient way to avoid having to say
anything more about the problem of fine tuning at all.

Both of these responses to the multiverse hoped it would help to
avoid metaphysics altogether. Alas, things are not so simple in the
multiverse. We have seen that once conscious observers are allowed
to intervene wilfully in the universe, rather than being merely lumped
into the category of passive 'observers' who do nothing except exist,
then we end up with a scenario in which the gods reappear in unlimited numbers in the guise of the simulators who have powers of life
and death over the simulated realities that they bring into being. The
simulators determine the laws, and can change the laws, that govern
their worlds. They can engineer anthropic fine tunings. They can pull
the plug on the simulation at any moment, intervene or distance
themselves from their simulation; watch as the simulated creatures
argue about whether there is a god who controls or intervenes; work
miracles or impose their ethical principles upon the simulated reality.
All the time they can avoid having even a twinge of conscience about

hurting anyone because their toy reality isn't real, is it? They can even watch their simulated realities grow to a level of sophistication that allows them to simulate higher-order realities of their own.

Faced with these perplexities, do we have any chance of winnowing fake realities from true? What might we expect to see if we are making scientific observations from within a simulated reality?

First, the simulators will have been tempted to avoid the complexity of using a fully consistent set of laws of Nature in their worlds when they can simply patch in 'realistic' effects. When the Disney Company makes a film that features the reflection of light from the surface of a lake, it does not use the laws of quantum electrodynamics and optics to compute the light scattering. That would require a stupendous amount of computing power and detail. Instead, the simulation of the light scattering is replaced by plausible rules of thumb that are much briefer than the real thing but give a realistic appearance – as long as no one looks too closely. There would be an economic and practical imperative for simulated realities to stay that way if they were purely for entertainment. But such limitations on the complexity of the simulation's programming would presumably cause occasional tell-tale problems – the 'creaking scenery problem' familiar from amateur dramatics – and perhaps they would even be visible from within.

Even if the simulators were scrupulous about simulating the laws of Nature, there would be limits to what they could do. While they might have a very advanced knowledge of the laws of Nature, it's unlikely they would have *complete* knowledge of them (some philosophers of science would argue this must always be the case). They may know a lot about the programming needed to simulate a universe but there will be some gaps or, worse still, errors in their knowledge of the laws of Nature. The lacunae would be subtle and far from obvious; otherwise our hypothetical 'advanced' civilisation wouldn't be advanced.

These shortcomings will not prevent simulations being created and running smoothly for long periods of time. But gradually the little flaws will begin to build up. Eventually, their effects will snowball and the virtual realities would 'crash'. The only escape is if their creators intervene to patch up the problems one by one as they arise. This is a solution that will be very familiar to the owner of any home computer who receives regular updates in order to protect it against new forms of viral invasion or repair programming gaps that its original creators had not foreseen. The creators of a simulation could offer this type of temporary protection, updating the working laws of Nature to include extra things they had learnt since the simulation was initiated.

In this kind of situation, logical contradictions will inevitably arise and the laws in the simulations will appear to break down now and again. The inhabitants of the simulation – especially the simulated scientists – will occasionally be puzzled by the experimental results they obtain. The simulated astronomers might, for instance, make observations that show that their so-called constants of Nature are very slowly changing.[31]

It's likely there could even be sudden glitches in the laws that govern these simulated realities. This is because the simulators would most likely use a technique that has been found effective in all other simulations of complex systems: the use of error-correcting codes to put things back on track.

Take our genetic code, for example. If it were left to its own devices we would not last very long. Errors would accumulate and harmful mutations and death would inescapably follow. We are protected from this degeneration by the existence of a mechanism for error correction that identifies and corrects mistakes in genetic coding. Many of our complex computer systems possess the same type of internal 'spell-checker' to guard against error accumulation, following the pioneering invention of error-correcting codes by Richard Hamming at Bell Laboratories in 1950.

If the simulators used error-correcting computer codes to guard against the fallibility of their simulations as a whole (as well as simulating them on a smaller scale in our genetic code), every so often a correction would take place to the state or the laws governing the simulation. Mysterious sudden changes would occur that seem to contravene the very laws of Nature that the simulated scientists were in the habit of observing and predicting.

We might also expect that simulated realities would possess a comparable level of maximum computational complexity across the board. The simulated creatures should have a similar complexity to the most complex simulated non-living structures – something for which Stephen Wolfram (for quite different reasons) has coined the term the 'Principle of Computational Equivalence'.[32]

One of the commonest worries about distinguishing a simulated reality from a true one from the inside is the suggestion that the simulators would be able to foresee some looming flaw ahead of time and pre-adjust the simulation to avoid the mismatch. This new simulated reality might then develop its own disparities with true reality but they could be anticipated by another act of predestination.

The question is whether this foresight is possible when taken to the limit. The problem is similar to one considered by Karl Popper to identify the self-referential limits of computers.[33] The same argument was used in a different context by the late Donald MacKay as an argument against the logical possibility of your future actions being predictable if the forecast is made known to you.[34]

It is only possible for me to make a correct prediction of all your future actions if it is not made known to you.[35] Once my prediction is made known to you, it is always possible for you to falsify it. Hence it is not possible for there to be an unconditionally binding prediction of your future actions. The same argument applies to more mundane things like predicting elections: there cannot be a public prediction of the outcome of an election that unconditionally takes into account the effect of the prediction itself on the electorate.[36] This type of uncertainty is irreducible in principle. Yet if the prediction is not made public it could be 100 per cent correct.

All this suggests that if we live in a simulated reality we should expect occasional sudden glitches, small drifts in the supposed constants and laws of Nature over time,[37] and a dawning realisation that Nature's flaws are as important as its laws for our understanding of true reality.

The point of this little excursion into territory usually occupied by science fiction writers is simply to show that unusual consequences seem to follow if we take seriously the idea that there exist an infinite number of possible worlds which fill out all possibilities. We can imagine how an extension of some of the science and technology we have at the moment would enable our successors to do some of these things. We don't even need to invent new science. Technological progress does the rest.[38] The implications for the nature of the universe that we see and its likely fallibility are striking, worrying even, and they take us back to the words of the philosopher David Hume at the end of the eighteenth century.

Hume's sceptical dialogues take many of the arguments for the existence of God that were fashionable at the time and pick on their presumptions about the perfect nature of creation, the uniqueness of the Deity and so forth.[39] Here is what he had to say about 'many worlds' and their likely defects:

You must acknowledge, that it is impossible for us to tell, from our limited views, whether this system contains any great faults, or deserves any considerable praise, if compared to other possible, and even real systems . . . If we survey a ship, what an exalted idea must we form of the ingenuity of the carpenter, who framed so complicated, useful and beautiful a machine? And what surprise must we entertain, when we find him a stupid mechanic, who imitated others, and copied an art, which, through a long succession of ages, after multiplied trials, mistakes, corrections, deliberations, and controversies, had been gradually improving?

Many worlds might have been botched and bungled, throughout an eternity, when this system was struck out: Much labour lost: Many fruitless trails made: And a slow, but continued improvement carried on during infinite ages in the art of world-making . . .

This world, for aught he knows, is faulty and imperfect, compared to a superior standard, and was only the first rude essay of some infant deity who afterwards abandoned it, ashamed of his lame performance; it is the work of some dependent, inferior deity; and is the object of derision to his superiors; it is the product of old age and dotage in some superannuated deity and ever since his death has run on at adventures, from the first impulse and active force which it received from him.[40]

Hume's scenarios conjured up images of a host of gods of varying degrees of competence creating universes of different quality, like apprentices attempting to copy the master. But if we replace his 'infant' and 'superannuated' deities by simulators, what he envisions is a realm where simulated universes abound: some good, some promising, others defective and doomed to be lifeless.

If all possible worlds exist and we are living in a simulation whose laws are not quite consistent with one another, does this make any difference? Indeed, should it make any difference?[41] It will be rather de-motivating if you are a (simulated) scientist trying to understand the way the world works. Anything could happen without reason.

Not surprisingly, simulated realities are not welcomed into the scientific world-view because they undermine it. Philosophers take them more seriously and some have even tried to use them as arenas to discuss ethics. The problems they spawn are unusual.

Hanson has suggested that the possibility of being in a simulated reality might produce its own influences on how you should act.[42] Simulated experiences, no matter how real they may seem, are much more likely to be brought to a sudden and unpredictable end than typical real experiences. This suggests to Hanson that 'all else being equal you should care less about the future of yourself and of humanity, and live more for today.' We are familiar with the fact that in films and the theatre the star is surrounded by other good actors who have to interact with the star, but as you move further away from the star then extras and low-paid jobbing actors can fill in the crowd scenes and non-speaking parts at low cost. Likewise, in a simulated reality, the characters far from your action may just be fake simulated characters and you shouldn't worry too much about them. Above all, Hanson suggests, if you are part of somebody's simulation, be entertaining! Be famous! Be a pivotal person! This will increase the chances of your simulated existence continuing and others will want to resimulate you in the future! Fail to have these characteristics and you could become like the soap opera character that quickly gets written out of the show, taking a long holiday to Vladivostok, never to return.

As we look around at the way people in the news do behave, and re-read Hanson's advice, we might easily be drawn to the conclusion that we must be living in a simulation. However, none of this is very persuasive. How you *should* behave depends entirely on the moral stance of the simulators. If they like to be entertained, you will do well to be entertaining. But if they are dedicated to a noble purpose, you might have the greatest chance of continuing re-creation and re-simulation by being a martyr for a just and good cause. While we do not suggest that these codes of behaviour are taken seriously as the basis for how to live your life, they do bring sharply into focus the central problems of moral philosophy and our responses to them. If simulated realities are common and we are in one of them, it would be worrying if they are simulations of the sort that we know. But why should they be? If we had always used the word 'simulation' to describe the result of a one-off act of creation by God, we are in a very similar situation, albeit with a simulator of a different sort.

These extraordinary consequences for life in simulated realities have led some to regard them as strong arguments *against* the existence of other worlds. If most of these worlds are virtual, they can display illusory laws of physics and we are on a slippery slope to knowing nothing at all because there is no reliable knowledge to be had. It is the counterpoint to solipsism and has many of the same paralysing consequences for any future thinking. If all possibilities are infinite and actual then reality contains rather more than we can bear.

UNIVERSES WHERE NOTHING IS ORIGINAL

. . . the librarian deduced that the Library is 'total' . . . and that its bookshelves contain all possible combinations of the twenty-two symbols. *All* – the detailed history of the future, the autobiographies of the archangels, the faithful catalog of the Library, thousands and thousands of false catalogs, the proof of the falsity of those false catalogs, a proof of the falsity of the *true* catalog, the Gnostic gospel of Basilides, the commentary upon that gospel, the commentary on the commentary on that gospel, the true story of your death, the translation of every book into every language, the interpolations of every book into all books, the treatise Bede could have written (but did not) on the mythology of the Saxon people, the lost books of Tacitus.
        Jorge Luis Borges[43]

Imagine living in a universe where nothing is original. Everything is a fake. No ideas are ever new. There is no novelty, no originality. Nothing is ever done for the first time and nothing will ever be done for the last time. Nothing is unique. Everyone possesses not just one double, but an unlimited number of them.

This unusual state of affairs exists if the universe is infinite in spatial extent and the probability that life can develop is not equal to zero. It occurs because of the remarkable way in which infinity is quite different from any large finite number, no matter how large that number might be.[44]

In a universe of infinite size and material extent, anything that has a non-zero probability of occurring somewhere must occur infinitely often.[45] At the present moment, there must be an infinite number of identical copies of each of us doing precisely what we are doing now. There are also infinite numbers of identical copies of each one of us doing something other than what we are doing at this moment. Indeed, an infinite number of copies of each of us are doing anything that it was possible for us to do with a non-zero probability at this moment. This alarming state of affairs is known as the 'replication paradox' and it was discussed by the German philosopher Friedrich Nietzsche in his book *The Will to Strength* in 1886, when these consequences of an infinite universe had been recognised. He writes:

the universe must go through a calculable number of combinations in the great game of chance which constitutes its existence . . . In infinity, at some moment or other, every possible combination must once have been realized; not only this, but it must also have been realized an infinite number of times.[46]

This replication paradox has all sorts of odd consequences. We believe that the evolution of life is possible with non-zero probability (because we are here). Hence, in an infinite universe there must exist an infinite number of living civilisations. Within them, there will exist copies of ourselves of all possible ages. When each of us dies there will always exist an infinite number of copies of ourselves elsewhere, possessing all the same memories and experiences of our past lives and who will live on to the future. This succession will continue indefinitely into the future and so in some sense each of us 'lives' for ever in this paradoxical scenario. One further odd conclusion arises here. If I consider all the possible histories in which I have my actual past but all possible futures, the number of them in which I cease to exist in the next few moments greatly outnumbers those in which I continue to exist. So why do I continue to exist?

This paradox also entered theological discussion in a provocative way. For suppose that we apply the same reasoning to the Incarnation. If it has a finite probability of occurring then it must have occurred infinitely often elsewhere in an infinitely large universe. This argument was used by St Augustine in the fourth century[47] to claim that sentient life must be unique to the Earth or the crucifixion would need to have occurred on infinitely many other worlds as well.[48] Thomas Paine, in the second half of the eighteenth century, argued that the existence of life elsewhere was obviously true and therefore the crucifixion did not occur (or at least could not have had its claimed effects) because it would be absurd for it to be infinitely replicated.

We could ask what might happen if we were to meet one of our copies. You might think this was just like shadow boxing in the mirror but there is no reason to think that any one of your doubles would act as you do. You may both have identical histories up until the moment of encounter, but confronted with a new situation you might respond differently, for the first time, just as two identical twins might do. In the future your experiences and choices would become increasingly different. Yet elsewhere in the infinite universe there would have to be a never-ending series of copies of each of us making the same future decisions and being in every respect identical. It is as if every possible decision that we could have taken at every moment is actually taken. There is always someone somewhere who lives a past life identical to my own and then takes one of all the possible decisions open to me about what to do next. They are always more numerous than those who continue to choose like I do.

One of the most curious features of this 'theory' is that, if it is true, it cannot be original. It has already been proposed infinitely often in the past. Indeed, in an infinite universe that explores all possibilities, nothing can be original.

Some cosmologists find these prospects so alarming that they would regard them as valid arguments for a finite universe.[49] Others are alarmed at the ethical consequences of a universe where all possible sequences of events take place regardless of their consequences. All possible virtual histories are acted out for real. There are histories where evil always overcomes good and others where that is even thought to be a good thing.

In reality, the laws of physics offer us a certain protection against some of these worrying conclusions. Even though the whole universe may be infinite we can only ever have contact with a finite part of it because the speed of light is finite. The distance to that contact horizon today is about 10^{27} metres. For comparison, the size of the Earth is 1.3×10^7 metres and the distance to the nearest star about 6×10^{16} metres.

These are huge distances but they pale into insignificance when compared with the distance you would have to travel in order to have a good chance of encountering a double. To have a good chance of running into your own double you would need to look out to $10^{10^{28}}$ metres and for a copy of the whole Earth and everything on it you would need to go to $10^{10^{30}}$ metres. The first copy of our entire visible universe would not be found with a 50:50 chance until we looked out to $10^{10^{120}}$ metres. The finiteness of the speed of light protects us from any encounters with our doubles in an infinite universe. Yet there is still something deeply disturbing about their existence despite our insulation from encountering them.

If an infinite universe has always existed in a roughly similar state, as in the old steady-state cosmology, there is also a temporal version of these paradoxes. Anything that has a finite chance of occurring will happen infinitely often in our past history. There can be no novelty in such a universe. Moreover, since there is a finite chance of intelligent life evolving, it must be infinitely common and as time goes on there should be a huge proliferation of living beings in the universe. We should expect to see them routinely. This odd conclusion and the mysterious absence of any evidence for extraterrestrials were used as an observational argument against steady-state and static cosmologies by Paul Davies[50] and by Frank Tipler and myself.[51] Of course, all these conclusions about infinite universes, if true, cannot be original.

Finally, a sobering and tantalising thought: we feel dismayed or sceptical about the very idea of infinite replication. It seems fantastic, ridiculous and impossible in equal measure. But all around us there are replicas that we habitually assume to be perfect. The world is made of them. Protons, electrons, quarks, all these elementary particles of Nature come in families of *identical* particles. Once you've seen one electron, you've seen them all.[52] No one knows why that is so. The universe is based upon replication and we suspect this replication will be infinite, just like the expanding universe appears to be. This is the most fantastic fine tuning of all. Most physicists don't even notice it and few ever comment on it. It suggests the deep-laid fabric of reality has replication at its heart.

BOLTZMANN'S UNIVERSE

'I believe we are on an irreversible trend toward more freedom
and democracy – but that could change.'
    Dan Quayle

The unsettling implications of infinite universes test our scientific
methods, as well as our credulity, to the limit. A different way to determine the likelihood of the existence of observers in the multiverse is
to ask for the probability that an observer would make some set of
observations about their universe. Unfortunately, that's not quite good
enough. If there is a finite chance of observers arising in an infinite
universe then they should exist in an infinite number of places, but the
things they will see will have different probabilities in each case. We are
prevented from making a definite prediction of the most likely set of
observations when there are infinitely many copies of observers.

Talk of 'observers' generally provokes the criticism that we are
hamstrung into thinking of life in too chauvinistic a fashion. We can
only imagine extraterrestrials that are a bit like us, made of atoms,
with brains, physical bodies; perhaps like computers, but none the
less just extrapolations of information-processors on Earth. Such
considerations are closely linked to the strategies of SETI, the search
for extra-terrestrial intelligence.[53]

In the past, the search for intelligent activity in the universe has
always focused on detecting radio signals or the signs of advanced
technological activity. The latter was expected to be messy and energy-
greedy, moving planets and influencing stars, and so be potentially
visible. In 1964, civilisations were even graded in terms of their energy
use by the Russian astronomer Nikolai Kardashev, according to
whether they could manipulate planets, or stars, or whole galaxies.[54]
I extended this to include a fourth type that could manipulate whole
universes in the ways we have seen in this chapter.[55] However, advanced
intelligence is more likely to make its technologies smaller. Our own
technologies have moved steadily towards increasing miniaturisation.
The study of quantum technologies has allowed us to manipulate individual atoms and construct machines using few atoms and molecules.
The exciting frontiers of nanotechnology and quantum computation

open up the possibility that huge quantities of information can be processed and stored in tiny volumes of space. Given the problems of environmental degradation created by advanced technological species, and the exhaustion of natural resources, we might expect that very advanced civilisations would have to move towards becoming technologically miniaturised. The state of advancement of a civilisation would be better gauged by its ability to engineer at smaller and smaller scales, rather than on larger and larger ones. Accordingly, one can classify civilisations according to their ability to manipulate molecules, atoms, elementary particles and space-time structure.[56] This means that the more advanced they become, the less visible are their technological activities because they use less energy and create smaller and smaller amounts of waste heat. Even their space probes might be no larger than clusters of atoms or molecules. We would not even notice them.

Cosmologists have contemplated even more extraordinary forms of 'intelligence' in the universe. As we have seen in chapter 2, in 1895, long before the idea of the expanding universe emerged, the great German physicist Ludwig Boltzmann started wrestling with the question of why we find disorder to increase with the passage of time (something we call the second law of thermodynamics, or the 'arrow of time'), whereas Newton's laws of motion always allow the time-reversed sequences of observed events to occur as well.[57]

Now, we recognise that there are indeed solutions to the laws of physics which describe wine glasses falling off the table and smashing into fragments on the floor and their time-reversed counterparts in which fragments coalesce into whole wine glasses. However, the starting conditions needed to realise the latter solutions are so improbable that we never see them. The time-reverse sequences in which disorder turns into order require very finely tuned starting conditions so that all the fragments are the right sizes and move in the right way so as to coalesce simultaneously and make a whole glass. It would be like dropping 10,000 jigsaw puzzle pieces and finding that the puzzle is correctly completed when the pieces hit the floor.

Boltzmann wondered whether different parts of the universe might have arisen, or evolved, into states with different degrees of thermodynamic improbability. A state with the same temperature everywhere is the most probable; any spontaneous deviation from this uniformity

250   *The Book of Universes*

is less probable. He imagined that overall, on the average, the universe was in thermal equilibrium. But for there to be life of any sort, there must be fluctuations where conditions deviate from equilibrium for various periods of time. In some of these fluctuations order will be seen to increase with time while in others it will tend to decrease, reflecting the degree of improbability of the starting state required when the fluctuation first occurred. Boltzmann argued with his student Ernst Zermelo (later to become a famous pure mathematician) that we must be living in one of the fairly long-lived fluctuations in which order tends to decrease with the passage of time.

Boltzmann offered two possible explanations for the tendency of disorder to increase. In one, the world was made in a highly ordered state so the most probable things that happen afterwards always go to more disorderly states (wine glasses will be seen to break into fragments). In the other, the universe contains all sorts of regions that began in different states of order and disorder and subsequently display different behaviours; some grow more orderly, some more chaotic, but life can only exist in one of the regions where disorder grows from an orderly beginning. Boltzmann imagined an unending collection of individual worlds, each formed as a rare fluctuation from equilibrium, separated by enormous distances – more than 10^{100} times the distance to a nearby star like Sirius. He avoided showing any preference for one scenario or the other and even gave his old laboratory assistant (Dr Schuetz) the credit for the idea of the second one.[58] As we saw in chapter 2, in fact the idea had been published twenty years earlier on several occasions[59] by the English engineer and physicist Samuel Tolver Preston, who came to Germany in order to study for a doctorate with Boltzmann, which he completed in 1894. I do wonder whether the elusive 'Dr Schuetz' was really Samuel Preston.

The second anthropic-fluctuation scenario attracted some interesting commentators. The influential French mathematician and physicist Henri Poincaré (1854–1912) liked it because it offered a possibility for humanity to avoid the ultimate heat death of the universe in a state of maximum entropy by inhabiting an atypical fluctuation where ultimate disorder was not inevitable. The British biologist J. B. S. Haldane thought that the extraordinary improbability of the right type of fluctuation lent support to the idea that life on Earth was unique. For if there were similar forms of life nearby, we would be faced with explaining why such a large, and therefore more improbable, fluctuation from equilibrium had arisen than was necessary to give rise to life on Earth. Taken a step further, Haldane's concern is really that we only need one life-supporting fluctuation, in which disorder increases, for there to be observers. It only needs to be as big as the solar system. This is vastly smaller – and so more probable – than the cosmic scales over which we observe the second law of thermodynamics to hold.[60] In the anthropic-fluctuation scenario it is highly improbable that the whole visible universe would display the same increase of disorder that we see on Earth.

There are two residues of this old thermodynamic argument in modern cosmology. The first is the argument of Roger Penrose that the universe is in an extremely ordered state today and therefore it must have begun 14 billion years ago in an even more fine-tuned orderly state.[61] This argument[62] has not proved very convincing to cosmologists because the measure chosen to gauge the disorder (or 'entropy') of the universe is just a measure of the age of the universe and any universe that is old enough for stars and carbon to form must have a very large entropy. More significantly, inflation is a way to generate local order by sweeping the disorder way beyond our visible horizon where we cannot see it. We don't know that the universe *as a whole* is in a low-entropy, ordered state. We only know what the visible part is like, so we can't conclude that the initial state of the entire universe was highly ordered or that the whole universe is in a very improbable state today.

Boltzmann's arguments have recently re-entered cosmology because his anthropic-fluctuation argument implies that if there are fluctuations from equilibrium, large enough and improbable enough to accommodate civilisations and minds like ours, the universe must be teeming with smaller, less ordered, disembodied, transient minds. They arise with greater probability than ourselves in this scenario and should arise at random infinitely often in an infinite universe.

These have been dubbed 'Boltzmann brains'. The paradox that such entities create for us is that it seems more likely that a Boltzmann brain arises randomly with false memories of a coherent life than that the universe around us would contain billions of beings like ourselves with self-aware brains and memories of complicated real experiences.[63] However, we have seen in this chapter that although the universe may be infinite, and also in some senses randomly infinite, there is more going on. There are laws governing how things change. There are physical processes that drive the expansion so fast that we are left seeing only a small and somewhat unusual part of a possibly infinite multiverse. The clear message of these developments, speculative as they may be, is that the fact of our own existence, and that of other conscious minds, no matter how unusual they may be, is a factor we must include in our considerations of what the universe may be like and how we test our theories. Observers are as important for theories of the universe as they are for the universe.

11 Fringe Universes

Theories, theories, myriads upon myriads of them, streamed over me like windborne leaves, like the contents of some titanic paper factory flung aloft by the storm, like dust-clouds in the hurricane advance of the mind. Gasping in this vast whirling aridity, I almost forgot that in every mote of it lay some few spores of organic truth, most often parched and dead but sometimes living, pregnant, significant.

Olaf Stapleton[1]

WRAP-AROUND UNIVERSES

It's a funny place Reading. Even its biggest advocates tend to talk about it in relation to its proximity to somewhere else. 'It's only 25 minutes by train from central London', the duty manager enthused. 'The coach shuttle to Heathrow takes 45 minutes!' said the waitress in the Mall coffee shop.

British Airways magazine[2]

If universes can appear from 'nothing' without violating the laws of physics, perhaps the creation of the universe might be describable using those laws. Traditionally, cosmologists who believed there was a beginning to the universe had been content to regard that as a place and time where the laws broke down. Those laws came into being at once with space and time and the physical universe. All our talk of 'other' universes and the multiverse tends to call into question this view. Increasingly, we see our universe as an outcome of the laws of Nature, perhaps one among many in a multiverse: a little local event in an eternal history. While this downgrades the status of the vast universe that we see around us, it elevates the status of the laws of Nature. They appear to be able to control and permit more than the one universe we observe. Throughout the twentieth century, cosmologists grew used to studying the different types of universe that emerged from Einstein's equations, but they expected that some special principle, or starting state, would pick out one that best described the actual universe. Now, unexpectedly, we find that there might be room for many, perhaps all, of these possible universes somewhere in the multiverse.

What sort of universe would be most likely to be created from nothing? All universes obeying Einstein's universe equations seem to have zero *energy* and zero *electric charge*. These are two of the three most important quantities that are conserved in all physical processes. Energy and charge can be moved around and redistributed, but when you total each of them up, adding positives and negatives, the final answer must always be the same. Things look a little different with respect to the third important conserved quantity, angular momentum. This is a measure of rotation and we have already seen it is not required to be zero by Einstein's equations in the same way as the energy and electric charge. Gödel found there could be rotating universes but we have not yet found any evidence for overall rotation of the universe. Nor do we expect to, for if inflation took place it would drive any pre-existing rotation down to unobservably small levels.[3]

In 1973, the American particle physicist Edward Tryon[4] attempted a speculation of this sort (which Pascual Jordan and George Gamow had wondered about before[5]). He suggested that the entire universe might be a virtual fluctuation from the quantum vacuum, governed by the Heisenberg Uncertainty Principle that limits our simultaneous knowledge of the position and momentum, or the time of occurrence and energy, of anything in Nature. On this picture he thought the universe might be 'simply one of those things that happens from time to time'. The energy–time form of the uncertainty principle requires the product of the lifetime of a fluctuation and its energy to be larger than Planck's quantum constant of Nature ($\Delta t \times \Delta E > h$).[6] This suggests that a zero-energy fluctuation – like our universe – could have an infinite lifetime. In fact, the most likely lifetime is only 10^{-43} of a second. It needs inflation to act upon these minuscule fleeting fluctuations if they are to grow astronomically big and old. Tryon couldn't explain why our universe had lived so long, and inflation would not be discovered for another eight years. All he could suggest to reconcile this picture with the rarity of large, long-lived vacuum fluctuations was a simple anthropic explanation for the very rare fluctuation because it is a necessary condition for observers:

The logic of the situation dictates, however, that observers always find themselves in universes capable of generating life, and such universes are impressively large. We could not have seen this universe if its expansion–contraction time had been less than the 10^{10} yr required for *Homo sapiens* to evolve.

If a more sophisticated version of this idea could be made to work we would need to think about the most likely properties of a created universe. Suppose it was a finite universe. There are many of them with all possible sizes and we have seen why we would have to be residing in a big one, although it might have started very small and undergone inflation. The other question we have to ask about any universe that appears like this is what would be its *topology*?

We have seen how Einstein's equations produce universes with spaces that can be curved or flat. The distribution of matter within the universe dictates the local shape of the space. This information doesn't tell you everything about the overall shape of the space though. One global property remains undetermined by Einstein's equations and has to be assumed to take some simple (or even complicated) form. That property is called the 'topology' of the universe and its importance was recognised immediately by Alexander Friedmann in 1922, when he discovered the first expanding universes with positively and negatively curved spaces.

Topology is different to geometry. In chapter 2 we encountered the idea of non-Euclidean geometries on curved surfaces. Let us expand the discussion there a little further. On a curved surface we can find the curvature by marking three different points A, B and C, and then joining them, A to B, B to C and C to A, by the shortest possible routes on the surface. With a flat surface this creates a triangle whose three interior angles add to 180 degrees. Draw it on a sheet of paper. Roll up the paper into a cylinder with the triangle on the outside. If you look at the triangle on the cylinder you will see that its sides are still straight lines and its three interior angles still add to 180 degrees.

Surprisingly, the cylinder is not a curved geometrical surface. What distinguishes it from the flat sheet is its overall topology. The effect of changing the topology in this way can be dramatic. In the case of the universe, it could have a flat Euclidean geometry that extends without limit in every direction: its volume would be infinite. But if each of the three directions is wound up like a cylinder to make a three-dimensional ring doughnut, the space will be flat everywhere but the volume of space will now be finite (Figure 11.1). This was why Friedmann was very careful not to say that universes that had flat or negatively curved geometry were infinite while those with positive curvature were finite in volume. He gave the example of the cylinder to show that rolling up the space in each direction can create a universe of finite volume even when the space is flat or negatively curved. This type of mathematical subtlety was familiar to Friedmann but would have been lost on most astronomers of the day. Although Einstein's theory has the beautiful property that the distribution of matter and energy in the universe determines the geometry of its space and the rate of flow of time locally everywhere, it does not fix the topology of space.[7]

If our universe had one of these wrap-around topologies, and there are many varieties other than the simple ring doughnut, then measurements of the curvature and expansion rate might make it look as if it is an 'open', infinite, ever-expanding universe, whereas it would in reality be finite. This possibility was examined by several cosmologists in the 1970s. In 1971, George Ellis, then at Cambridge, investigated the catalogue of possible topologies[8] that the universe could have and found that some would create odd conflicts with aspects of physics – for example, in some, where space was given a twist before 'glueing' the opposite sides together, a right-handed elementary particle might go on a trip and return as a left-handed one – but there were plenty of others that avoided any of these problems.[9]

In 1974, two Soviet scientists, Dimitri Sokolov and Victor Shvartsman, focused on the observational consequences for astronomers of living in one of these wrap-around universes.[10] They started thinking about

Figure 11.1 A two-dimensional universe of space with the topology of the surface of a ring doughnut.

the consequences of the wrap for images of bright galaxies (Figure 11.2). It is as if we are living in a hall of mirrors. In each direction we see multiple images that seem to get fainter and fainter because the light has travelled more and more circuits of the finite universe to reach us. If you are in a room whose walls are mirrors you see an innumerable number of reflected images of yourself in every direction, each getting smaller and smaller as it fades into the distance.

The easiest way to see whether we live in a wrap-around universe is to pick on a big bright cluster of galaxies – the great Coma Cluster containing more than 1000 galaxies located 321 million light years away is a favourite – and look for copies of it in the sky. They won't be identical copies though because the images will picture the cluster at

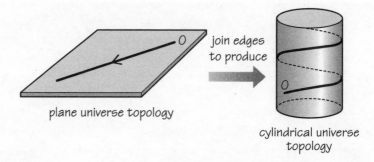

Figure 11.2 A cylindrical topology created by wrapping up a plane. An observer in the cylindrical universe receives photons that have travelled a long time without coming from so far away as in the plane universe and they have circled the space many times.

different times in its life and the light will have followed different paths en route to us. None the less, the first copy should be obvious if mirror imaging is occurring.

These considerations were able to place a limit on how small the universe could be, if it was using a wrap-around topology to create the illusion of being infinitely large. The analysis of Sokolov and Shvartsman, and a subsequent re-examination by Richard Gott in 1980,[11] found a lower limit of between 400 million and 600 million parsecs[12] on the size of the whole universe. At this time the inflationary universe had not yet been discovered and the prevailing view was that there was a single 'universe' which might be either finite or infinite.

In 1984, Yakob Zeldovich and Andrei Starobinsky in Moscow returned to the question of creating universes out of 'nothing' – or at least out of the quantum vacuum – if they had different possible wrap-around topologies.[13] At that time it was thought that only finite universes could be created in this way[14] so it was interesting to discover if the unusual topologies were more likely to arise than the standard one. They found that a simple flat universe with a finite wrap-around topology could appear as a quantum fluctuation so long as the wrap-around distances in the three different directions were roughly the same. If they were very different in size in each direction, the universe must emerge from a singularity of infinite density.

There is no way to decide what the topology of the universe

should be from Einstein's equations alone. Traditionally, cosmologists have adopted the simplest possible topologies but it could be argued that there are many more exotic wrap-around ones than simple ones, so the wrap-around universes might be more likely if we were picking universes at random out of a hat. Others argue that finite universes are more natural and avoid alarming consequences, like the replication paradoxes, and also permit quantum theory to work more naturally in the universe. This means that all the finite universes need to be considered, not just the simplest ones with spherical topology that Friedmann first examined. The opponents of wrap-around universes are sceptical. Why should the scale of identification of space be in an interesting range? It would seem to be a fantastic coincidence if it was just observable today. This objection is a good one if the universe has zero curvature and space is precisely Euclidean. There is then no reason why the wrap-around scale should be anything near the scale of our visible horizon. It could be anything. But if the universe is negatively curved, it has a curvature radius that is very close to the size of the visible universe today and the wrap-around scale might be closely related to it, which would be far from a coincidence.

This issue of topology has been resurrected in recent years as astronomers try to account for every detail in the cosmic microwave background radiation. At first, there seemed to be a slight deficit in the variations in temperature over about 10 degree separations on the sky. This might be explained by a wrap-around universe because the large waves of temperature variation would not easily 'fit' in the 'small' finite universe and the temperature fluctuations would be diminished near the wrap-around scale.[15] However, recently there has been evidence that some of these reported deficits might be insignificant statistically. We will no doubt hear more about this in future years when the Planck satellite reports on its sky survey in 2011.

QUANTUM UNIVERSES

'The number you have dialled is imaginary. Please rotate your
phone through ninety degrees and dial again.'
    'Imaginary' answer-phone message

One of the greatest challenges of modern physics has been to join
together Einstein's theory of gravity and the quantum theory of matter
and light. Traditionally, quantum theory governs the small-scale world
of atoms and their constituents where gravity is too weak to be noticed.
Gravitation governs the largest structures in the universe today. But if
we go backwards in time towards the apparent beginning of the expansion of the universe, we encounter times when we know that gravity
and quantum theory must join together to form a new environment
where the uncertainties of quantum reality carry over into the fabric
of space and time themselves. Remarkably, the constants of Nature
uniquely define the time when this breakdown must happen.

The lesson we learn from quantum theory is that all particles have
a quantum wavelike aspect. The wavelengths of their quantum effects
are inversely proportional to their masses. When the quantum wavelength exceeds the physical size of a particle it behaves in an intrinsically
quantum wavy fashion. In the opposite situation, as is the case for you
and me when we walk down the street, the quantum waviness is negligible. If we follow the mass of the part of the universe that is contained
within a ball whose radius is equal to the speed of light times the age
of the universe, we can ask when the quantum wavelength of that mass
exceeds the diameter of the ball. In that situation the whole causally
connected region of the universe will behave like a quantum wave and
we expect Einstein's universe equations to fail. This watershed time is
very small, and is equal to $t_Q = 10^{-43}$ seconds. In that time light signals
will only have been able to travel about 10^{-33} centimetres.

This smallest 'tick' of time is very fundamental. It is the natural
unit of time defined by the laws of physics with no anthropomorphic
bias. It is defined by the constants of nature capturing the quantum,
relativistic and gravitational qualities of the universe alone.[16] It seems
so small because we are measuring it in 'human' units – seconds – that
characterise distinguishable instants of our experience. We say the

universe is about 14 billion years old; that is, about 10^{60} of these quantum ticks. In this sense the universe is very old. A young universe would have an age of just a few quantum ticks.

We said that when the universe is one 'tick' old the size of the ball across which light signals can travel is about 10^{-33} centimetres. This seems unimaginably small but there is a good way to bring it within our grasp. Take a sheet of A4 paper and imagine slicing it in half, again and again. After you have halved it just 30 times it will be the size of a single atom. Halve it 47 times and it will be the size of a single proton. Halve it 114 times and it will be about 10^{-33} centimetres across. Just 114 halves take us from a sheet of A4 to the smallest dimension over which the notion of distance still makes any physical sense. Going in the other direction, if you double the size of the paper just 90 times you will be up to the scale of the entire observable universe today – about 14 billion light years. So these unimaginable distances, both large and small, are really just paper tigers.

We notice that the universe is not expected to start inflating until it is about 100 million quantum 'ticks' old, at $t = 10^{-35}$ seconds, so the problems of quantum gravity don't affect those events. However, if we want to probe what the universe was like before inflation and ask whether it had a beginning in time, we have to confront the problem of quantum gravity head on.

We have seen that the old mathematical studies of Einstein's equations pioneered by Roger Penrose and Stephen Hawking in the mid-1960s spelt out very precise conditions under which the history of the universe experienced a beginning in the past. Unfortunately, the assumptions from which this conclusion follows no longer hold when we extrapolate back into the quantum gravity era. Einstein's equations will have to be modified and gravity may no longer be attractive. This doesn't mean there wasn't a beginning if there was inflation but we can no longer deduce that there *must* have been one. Our theorems can't help us any more. In fact, the type of matter needed for inflation to occur and, as we shall see in the next chapter, to explain the observed expansion of the universe today must violate the assumptions needed to prove there was a beginning in time.

We have also seen that the eternal inflationary universe introduces a completely new perspective on the issue of the beginning of the universe as a whole. Each region, like the one we call our visible

Figure 11.3 John A. Wheeler (1911–2008).

universe, has a beginning in this theory, but the overall self-reproduction process which spawns countless inflating universes within the infinite multiverse need have no beginning. It may have always existed and will always exist in the future if the expectations of this theory are to be believed.

There is no agreed formulation of a quantum cosmology. In many ways the very idea is problematic because we are used to using quantum mechanics to predict what an observer will see when a measurement is made. For example, this might be to determine how many electrons emerge from radioactive decays in a particular energy range. Before the measurement everything is a wave of different possibilities; after the measurement a definite outcome is recorded. All we can do is predict the probability of any possible outcome being measured. However, when it comes to the universe there are no outside observers to make measurements of it and the whole philosophy of quantum mechanics changes. We must predict correlations between different things. For instance, if we observe the expansion rate of the universe to take some value today, what is the probability that we should also measure the clustering of galaxies to extend out to some

Figure 11.4 Bryce and Cecile DeWitt hiking in the French Alps in 1963.

distance scale? However, in practice we are far from being able to deduce any of those probabilities.

Many attempts to study the quantum mechanics of universes make use of a special equation found by John A. Wheeler and Bryce DeWitt in 1967, known as the 'Wheeler–DeWitt equation'.[17] (DeWitt always described it as the Einstein–Schrödinger equation and attributed it to Wheeler, while Wheeler insisted on calling it the DeWitt equation. They finally both agreed to call it the Wheeler–DeWitt equation in 1988.)

The Wheeler–DeWitt equation makes a first attempt at combining Einstein's equations of general relativity with the Schrödinger equation that describes how the quantum wave function changes with space and time. Its solutions give a wave function for the universe. If this equation could be solved it would tell us the probability that a universe would evolve from one state into another. In order to find solutions like this it is necessary to specify some starting conditions

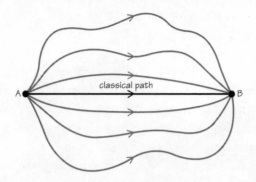

Figure 11.5 Evolutionary paths between two states of the universe, A and B. Newton's laws dictate that the single 'classical path' is followed. Quantum mechanics gives only the probability of there being a transition from A to B which is a weighted average over all the possible paths between A and B, some of which are shown here.

for the wave function of the universe and it is not obvious what they should be. Several explorations of this quandary have been made which serve to illustrate some of the peculiar consequences that a quantum universe might possess.

The approach of James Hartle and Stephen Hawking to the problem was very radical.[18] Hawking had outlined the general idea in a talk at the Pontifical Academy in the Vatican in 1982.[19] There seems to have been an intended resonance here with the Vatican's special interest in the 'beginning' of the universe, and Georges Lemaître, the architect of the big-bang universe idea, was the President of the Pontifical Academy from 1960 to 1966. Hartle and Hawking used an elegant formulation of quantum mechanics introduced by Richard Feynman to calculate the probability that the universe would be found to be in a particular state. In order to work out the probability that the universe will evolve from state A into state B you consider the contributions from all the possible paths through space and time that the evolution could take to get from A to B. In quantum mechanics the universe has some probability for taking any of them but in the situation where the universe gets bigger and older and quantum effects are less and less important one of these paths is dominant and the others cancel each other out like pairs of waves with their peaks and troughs super-imposed. This predominant evolution is called the 'classical path' and corresponds to what Einstein, or even Newton, would have predicted without including quantum mechanics (Figure 11.5).

Usually, when these calculations were done, the possible paths for evolution from A to B only included ones through space and time, called 'Lorentzian' paths. These are the ordinary histories that particles trace out if they move at the speed of light or slower. If you walk, or cycle, to work you follow a Lorentzian path through space and time. But Hartle and Hawking wanted to include another set of paths, called 'Euclidean' paths in which time was changed into a fourth dimension of space. This sounds odd but physicists had used the trick of converting time into space when beginning these calculations because it made them much easier. At the end they would change one of the space dimensions back into time again. It was just a handy method, like using different co-ordinates to plot a graph. But Hartle and Hawking didn't want to regard this as merely a handy aid to calculation: they proposed an initial state in which time had become another dimension of space.

This sounds very strange. But perhaps we can do without time? All the information you need to label different states of the universe and distinguish the future from the past can be provided without ever mentioning 'time'. Even today we notice something similar. If we used the temperature of the background radiation as a clock then the falling temperature could be used to distinguish the future from the past. In Hartle and Hawking's quantum universe the evolution of the universe from one state to another is dominated by the Euclidean paths when the universe is very hot and small but by the Lorentzian ones when it is big and cool.

There are two consequences that are striking. Time is not funda-mental in this theory. It is a quality that emerges when the universe gets large enough for the distinctive quantum effects to become negli-gible: time is something that arises concretely only in the limiting non-quantum environment. As we follow the Hartle–Hawking universe back to small sizes it becomes dominated by the Euclidean quantum paths. The concept of time disappears and the universe becomes increasingly like a four-dimensional space. There is no beginning to the universe *in time* because time disappears.

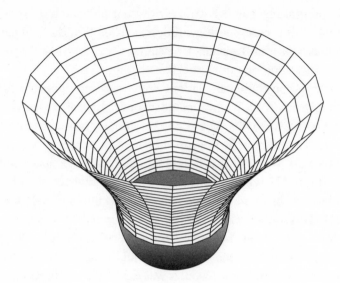

Figure 11.6 In the 'no boundary' proposal of James Hartle and Stephen Hawking for a cosmological initial state, time becomes a fourth dimension of space (or, so called, 'imaginary' time) when the universe is small. The up–down time axis rotates horizontally in the picture to create a smooth rounded boundary and time has no beginning. The space-time evolution of the early universe looks like a shuttlecock.

This approach is often described as making time 'imaginary' because the transformation that changes time into another dimension of space corresponds to multiplying it by an imaginary number, the square root of minus one. If you are not a mathematician this sounds exotic but it has a very simple geometrical interpretation. It corresponds to rotating the time axis on our graph by 90 degrees so that it becomes a space axis. If we look at our simple space-time diagram with light rays coming in towards us along straight lines from the past we can see what a switch to imaginary time will do in the past. The vertical time axis rotates around to become horizontal like space and the past becomes a smooth rounded surface like the bottom of a bottle (Figure 11.6).

This is the situation that Hartle and Hawking called the 'no boundary' state for the origin of the universe and it was the centre-piece of Hawking's best-selling book *A Brief History of Time*, published in 1988. It characterised the beginning of the universe as an event which did not play the role of a boundary in time. The universe has a beginning but not at a big-bang singularity of infinite density and temperature where space and time are destroyed. Its beginning is smooth and unremarkable, just like walking past the North Pole on the Earth's surface.[20] In effect, the no-boundary condition is a proposal for the state of the universe if it appears from nothing[21] in a quantum event. The story of this universe is that once upon a time there was no time.

This prescription is not without problems and is not the only possibility for the initial quantum state of the universe. Alex Vilenkin proposed another initial state which made very different predictions about the most likely type of universe to appear from nothing.[22] Vilenkin's starting condition eventually looked more reasonable because it favoured the appearance of universes that were very small with very high temperature and density – rather like the hot early universe that was the standard picture of the early universe. The Hartle–Hawking prescription, on the other hand, was found to require that the most probable universe was infinitely large and empty[23].

A SELF-CREATING UNIVERSE

'The only new thing in the world is the history you don't know.'
Harry S. Truman

The idea of a universe that doesn't have a hot, singular beginning because the notion of time fades away as we peer deeper and deeper into the past can also be realised without appealing to the subtleties of quantum mechanics and imaginary time. One simple possibility, which I suggested in 1986, was that there could be a universe with a past but no beginning by requiring all paths through space and time to be very large closed loops.[24]

The existence of closed loops which permit time travel is allowed by Einstein's equations, as we learnt from Gödel's non-expanding universe. It was non-expanding, but imagine there is an expanding universe in which all the histories of particles and light rays are closed loops of enormous (more than 100 billion years) total duration. We

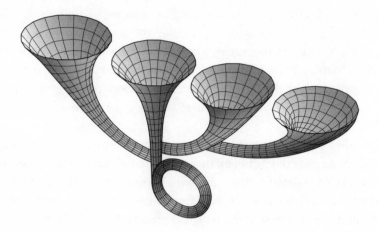

Figure 11.7 Gott's 'self-creating' universe. A loop in time at the beginning allows this universe to be its own progenitor. It is a non-quantum, real-time counterpart to the Hartle–Hawking prescription.

would not yet have noticed anything odd in this universe, but delving back into the past would eventually simply drop us down in the future. Such a universe doesn't have a beginning in time. It just 'is', although, to echo Bill Clinton, you do have to clarify what 'is' means.

This way of evading time's origin was developed in more detail later by Richard Gott and Li-Xin Li at Princeton University.[25] They modified the eternal inflationary universe theory to give an example of a universe that could be said to create itself.

The eternal inflationary universe sees new 'baby' universes being produced from a parent universe all the time. If we are in one of the universes in this sequence, we could track back and locate our 'mother' universe, and its mother universe, and so on. We have already seen that it is possible – some would argue most likely – that the tracking back never ends and there is no beginning to the universe-making process or the multiverse that defines it. But the new possibility suggested by Gott and Li is that in the past one of the branches loops back upon itself, producing a closed time loop, and so appears to 'create' itself. If all the branches emanate from one or more such incestuous branches, the universe would become its own mother and there would be no beginning to find (Figure 11.7).

COLLIDING UNIVERSES

On the M5 recently I was overtaken by a lorry thundering north
marked *Malvern Water*. A few minutes later I saw a similar vehicle
going south marked *Highland Spring Water*. Are we mad?
      Stephen Pimenoff[26]

The idea that expanding universes containing more than the critical
density might contract back to a big crunch and then 'bounce' into a
new expanding cycle which subsequently contracts and bounces, and
so on *ad infinitum*, is an old one. It was first suggested as an extension
of the simple 'one shot' closed-universe model by Richard Tolman in
1934, but he pointed out that the second law of thermodynamics would
lead to a systematic shift of energy into more disordered forms, like
radiation, which would cause successive cycles to grow in size and
age. More recently, in 1995, Mariusz Dąbrowski and I showed that if
Einstein's cosmological constant exists, then, no matter how small its
value, there will always be a last oscillation and the next state will be
converted into one of exponential expansion (see Figure 3.23).[27]

In 2001, interest was renewed in cyclic universes by a variant that
exploited new possibilities offered by string theory: it was proposed
by Justin Khoury, Burt Ovrut, Paul Steinhardt and Neil Turok.[28] They
named it the 'ekpyrotic' universe in celebration of the early Greek
Stoic philosophers, who favoured a cyclic view of the cosmos. The
Greek word *ekpyrosis* denotes the all-consuming fire into which they
believed the universe periodically plunged before emerging, phoenix-
like, into a new creation.

The new version of a cyclic universe proposed that there is a natural
initial state for the universe which is dictated by maximum symmetry.
String theories produce self-consistent theories of everything only if
the universe possesses more dimensions of space than the three we
are familiar with. In order to square this prediction with experience
it is believed that only three of the dimensions became very large,
perhaps by cosmological inflation acting selectively upon them in
some peculiar way, while the others remain imperceptibly small today.
This separation of dimensions into large and small must have happened
very early in the history of the universe close to 10^{-43}s. The proponents

of the ekpyrotic theory proposed that there is a natural initial state in which two three-dimensional universes (called 'braneworlds'[29]) approach each other by moving in one of those extra dimensions. Their motion is like two perfectly parallel sheets of energy approaching and colliding. When they collide they will produce a conflagration that rebounds back into a state of expansion. It is claimed that this collision and rebound does not suffer from the intractabilities of the traditional big bang because there are no infinities in physical quantities, like temperature and density, and the fabric of space and time continues smoothly. The energy released by the collision goes into creating elementary particles that participate in the expansion. The hope is that the resulting expansion will occur at the critical rate and will manifest small fluctuations that will show up in the temperature maps of the universe that satellites are completing now.

Can this rather adventurous theory predict anything that is both observable and different from the predictions of inflation? Is the theory consistent? These are key questions that its proponents are still grappling with. It offers a scenario which may just have one collision that converts contraction into expansion but it also permits an indefinite sequence of expansions and recontractions.[30] The total entropy of the whole universe increases from cycle to cycle but the amount that you see in the observable part of the universe doesn't grow unacceptably large because the rapid accelerated expansion early in each cycle dilutes the entropy produced in the last one. But there is still a similar problem with the accumulation of long-lived black holes from cycle to cycle which get caught up in the next collision of worlds, and with the rapid growth of anisotropies in the expansion each time the contraction occurs.[31]

Other versions of speculative high-energy physics have incorporated considerations of braneworlds, but without cataclysmic collisions. There could exist another three-dimensional universe that lies very close to ours when gauged in one of the extra dimensions of space. If the other brane moves with respect to us, this can create modifications in observable aspects of physics: traditional constants of Nature might change and the inflation of the universe might be stimulated. The motivation for considering schemes like this is not primarily cosmological though. It was proposed that the force of gravity acts in all the dimensions of space but the other strong, weak and electromagnetic forces do not (see Figure 11.8). This may offer some clue as to why

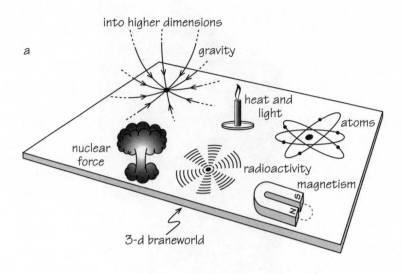

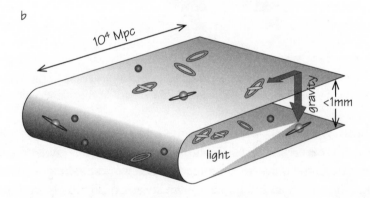

Figure 11.8 (a) The forces of electromagnetism, radioactivity and nuclear interactions are confined to a three-dimensional braneworld while gravity acts in all the dimensions of space, and so is found to be much weaker than other forces on the braneworlds. (b) Branes can be parallel and separated by a distance of less than a millimetre but they can also be folded over and separated by equally small distances. Light travels around the folded surface of space on the brane but gravity acts in higher dimension as well, and can propagate across the 'gap' between the folds creating the appearance of much stronger gravity over distances smaller than the brane separation.

Figure 11.9 Our universe could be a brane located close to another parallel braneworld. Both are located on the surface of the overall 'bulk' space.

gravity is so much weaker than the other forces. And even, perhaps, why there seems to be far less luminous matter than gravitating matter on view in the universe. Our nearby braneworld can be gravitationally close but optically distant, as shown schematically in Figure 11.9.

When we looked at the chaotic and eternal inflationary universe pictures we tacitly assumed that our patch had inflated a lot, vastly more than the extent of our visible horizon today. It is possible, although rather unnatural, that if the amount of inflation was only (coincidentally) enough to explain what we see, we could already be encountering the effects of the different sort of inflation that occurred 'next door'. If so, we could be looking for effects of a neighbouring bubble impinging upon our own expansion. There would be major effects eventually that would totally distort the expansion we see and the isotropy of the background radiation. But what about the effects of first contact with

Figure 11.10 Temperature bands that would be created on the microwave background radiation over the sky by two bubble 'universe' glancing collisions.

an alien bubble? Some attempts have been made to predict what we might see if a gently glancing bubble were to leave some 'scars' at the edge of our universe. We would look for stripes in the temperature pattern of the radiation over the sky that this would be likely to give rise to.[32] A simulation of the effects has been made by Michael Salem and the stripy effect on the microwave background radiation is illustrated in Figure 11.10. This is only for the simplest scenario though; with more complicated impinging bubbles there might be very different physics (or antimatter only) that would produce catastrophic effects on contact with our own bubble. Fortunately, we expect universes going bump in the night to be fantastically unlikely, but we once thought collisions of the Earth with asteroids and comets were unlikely too!

THE DYING OF THE LIGHT

> Do not go gentle into that good night.
> Rage, rage against the dying of the light.
>   Dylan Thomas[33]

We have already seen that on several occasions cosmologists have explored the possibility that traditional constants of Nature are changing with time, or might vary from place to place in the universe. The most radical possibility of this sort was developed in 1998 by Andy Albrecht, João Magueijo and myself.[34] It explored what would happen if the speed of light were to change in the early stages of the universe. This idea turned out to have some very interesting consequences that were just like those arising from a brief bout of inflation or from accelerated expansion. If there was a short period in the very early history of the universe where the speed of light fell in value, the expansion is driven towards the critical divide separating open from closed universes, monopoles are removed and irregularities ironed out.[35] All this is very desirable and offers another solution for the variety of problems that confronted cosmologists in the 1980s. It also seems to reduce the influence of the cosmological constant term in the late stages of the expansion, something that inflation cannot do.[36]

This unusual scenario does not require any residual change in the speed of light today. The change in light speed needed to produce these results could have occurred fleetingly, like inflation, when the universe was a mere 10^{-35} seconds old. The challenge it faced was to find a way to generate fluctuations in the density of the universe that could form the seeds of galaxies and reveal themselves by the temperature fluctuations they create in the background radiation. Inflation was able to do both those things and looked more appealing. Attempts have been made to generate fluctuations in other ways in cosmological theories like these but they remain to be fully explored.[37]

Schemes in which the speed of light varies are known as VSL ('varying speed of light') cosmologies. They show how open cosmologists have become to the idea that fundamental constants, once sacrosanct, might be varying. One reason for this liberalisation of approach is the way in which many constants have been seen to arise in string theories. The vast multiplicity of vacuum states found in those theories means that there are countless different 'deals' of the constants which produce self-consistent possible universes. Constants look like ordinary things that can fall out with all sorts of different values and can change if the universe evolves from one vacuum to another. Their status in the grand scheme of things has been significantly lowered. This change of attitude has been reinforced by another new perspective on the universe which downgrades the status of our 'constants' even further. It invites us to look for the true constants of

Nature outside the dimensions of space and time that we experience directly.

HYPERUNIVERSES

'But do you really mean, Sir,' said Peter, 'that there could be other worlds – all over the place, just round the corner – like that?'

'Nothing is more probable,' said the Professor, taking off his spectacles and beginning to polish them, while he muttered to himself, 'I wonder what they *do* teach them at these schools.'

C. S. Lewis, *The Lion, the Witch and the Wardrobe*

The quest for a new Theory of Everything led the string theorists to a remarkable early discovery: such unified theories were only possible if there were many more dimensions of space than the three we are familiar with in everyday life. The introduction of the idea that there are more than three dimensions has opened up many new possibilities. In fact, string theories and the deeper 'M theory' allow there to be more than one dimension of time. These theories only exist if the total dimension of space *and* time is given by a particular value, typically 10 or 11. This is usually interpreted as meaning that there are nine or ten dimensions of space and one of time. However, that split between space and time is not specified by the theory: they could exist just as well if there were three time dimensions and seven or eight space dimensions. We just assume there is only one time because things get very strange with more than one time: unstable particles decay very quickly, energy disappears, and the future is not uniquely and completely determined by the present. It's strange but not logically impossible or physically inconsistent. Very likely, it would be impossible for complex life to evolve in a two-timing universe but maybe other times can be subtly innocuous by being very brief in extent – just tiny little extra space dimensions.

It could be that there are logically possible universes with different space-time splittings populating the landscape of possible vacuum states for the universe. The outcomes of inflation might also contain different possibilities for the split between space and time, as well as for the number of dimensions that inflate and become large. If so, we suspect that we would have to find ourselves in one with a 1 + 3 split of time and space dimensions. In both cases, the choices that fix the number of times and spaces, and the number of space dimensions which become large, might be determined completely at random or dictated by some as yet unknown principle.

The most sobering feature of these developments in fundamental physics that make us take extra dimensions of space seriously is that the three-dimensional space around us that we call 'the universe' is just a shadow of a truer higher-dimensional reality. All the quantities that we called 'the constants of nature', for example, are not truly fundamental at all. The true constants exist in nine or ten dimensions and we just see part of them projected into our three dimensions. As a result, our so-called 'constants' need not even be constant. If the extra dimensions wobble, or very slowly change in size, we would see changes occurring in our three-dimensional constants at the same rate. This is why there is tremendous interest in the observational search for possible variations in some of the constants of Nature using astronomy and precision laboratory experiments.[38] There is even a growing body of data from quasars that is strongly consistent with a tiny variation in the constant determining the strength of electricity, by a few parts in a million over 10 billion years.[39] Variations in the traditional constants of Nature are a powerful window through which to look into extra dimensions. Other physicists have hopes that events detected in the Large Hadron Collider might provide evidence for or against the idea that there is another braneworld universe very close by into which particles can decay and leave the tell-tale trace of energy mysteriously disappearing from space. The most interesting thing about these speculative possibilities is that they are not unbridled: they are amenable to experimental constraint and test using particle accelerators and powerful astronomical telescopes on Earth and in space.

12 The Runaway Universe

'The universe may be like Los Angeles.
It's one-third substance and two-thirds energy.'
　　Robert Kirshner[1]

THE BEST-BUY UNIVERSE

The sane person prides himself on his ability to be unaffected
by important facts, and interested in unimportant ones. He refers
to this as having a sense of perspective, or keeping things 'in
proportion' . . .
　　Celia Green[2]

In the summer of 1996 a large cosmology conference was held at
Princeton in New Jersey as part of the 250th anniversary of the
founding of Princeton University.[3] Weather conditions – heat, swel-
tering humidity and thunderstorms – were appalling and not helped
by imperceptible air conditioning in the old student rooms that the
participants found themselves staying in. It was a relief to get to the
auditorium. One of the novelties of this conference was that it did
not just offer the usual spectrum of talks, but pairs and trios of speakers
were set up like political candidates to 'sell' you a particular model
of the universe in preference to that offered by their rivals. After
making their sales pitches the speakers entered into a critical debate
with each other in which the audience joined in.

Generally speaking, cosmologists at that time were content with
the idea of the inflationary universe; they didn't think too much about
chaotic or eternal inflation, and the multiverse was not a word that

had entered their vocabulary, although the concept was familiar from past anthropic arguments. The conference was more engaged with the detailed observations about the rate of expansion of the universe, its age, whether galaxies could form in time, and whether the pattern of non-uniformities in matter and radiation in the universe agreed with any of the theories about the origin of irregularities from inflation in the very early universe.

The challenge for each cosmological speaker was to show that their model for the matter content of the universe and the type of expansion was the best fit to all the observations. The front runner turned out to be the model, advocated by Michael Turner, which was expanding almost at the critical rate, just as inflation predicted, but which possessed a small positive value of the notorious cosmological constant that Einstein had invented and then rejected, with its repulsive gravitational effect accelerating the universe today. As Turner pointed out, the success of this model over its rivals was not exactly a surprise because it was largely the same as its competitors but with an added quantity (the cosmological constant) that could be tweaked so as to fit the observations a little bit better.

This winning model had been given the name 'Lambda-CDM': 'Lambda' describes the cosmological constant while 'CDM' is the abbreviation for 'cold dark matter'. This was a label for the form of matter that was required in all descriptions of the universe in order to account for the fact that the abundance of luminous matter was observed to be ten times too small to account for the strengths of the gravitational fields displayed in galaxies and clusters. There had to be lots of dark matter to explain this mismatch. It had to reside in a special form which only takes part in gravitational or weak interactions; otherwise it would suppress the production of deuterium nuclei below what was observed when the universe was three minutes old. This meant that the dark matter most likely had to be a neutrino or new type of neutrino-like particle, which felt the weak interaction. The known types of neutrino didn't fit the bill. They were too light-weight and created the wrong patterns of clustering when their behaviour was first explored in 1985 using big computer simulations of the expanding universe.[4]

In order to meet all these requirements the neutrino-like particles had to be much heavier than protons and therefore very slow moving, hence the label 'cold' – temperature is just a name we use for the average speed of molecules in a gas. Their sluggish behaviour created a distinctive type of small-scale galaxy clustering in the computer simulations that was a very good match to observations. Lambda-CDM, with the extra lambda parameter added, came out just ahead on all the astronomical scorecards. But nobody got excited about the success of Lambda-CDM, not even its advocate: it just looked so contrived and, to be honest, plain ugly.

This best-fit universe was much the same as Lemaître's universe from sixty years before. Like Einstein before them, cosmologists of the day had lost interest in the cosmological constant. It needed a fantastically small value (10^{-120}) in order to play its required role in the best-fit universe. This value was so small that many physicists thought that it was just a signal that its true value was *zero*: there was a deep principle of physics waiting to be discovered that forced its value to be precisely nought. One day we would find that new symmetry principle. Until then just ignore it. That was a common attitude amongst particle physicists. Astronomers, meanwhile, were always sceptical about parts of their data. It was so easy for the case for lambda to rest on something that would eventually fade away or be found to be more uncertain than we first thought. Even those who took Lambda-CDM more seriously were somewhat cautious because the evidence for its presence wasn't direct. We weren't observing the acceleration of the universe directly today, just catching small corrections that it contributed to other observations about the universe's behaviour in the past.

Things changed quite dramatically in 1998. Two large research teams, working independently and led by world-class astronomers, found the first direct evidence that the expansion of the universe is accelerating *now*. To everyone's surprise the High-z Supernova Project led by Adam Riess at Harvard University and the Supernova Cosmology Project led by Saul Perlmutter at the Lawrence Berkeley Laboratories at the University of California at Berkeley both found spectacular new evidence that the expansion of the universe began accelerating just a few billion years ago.[5] You needed to extend Hubble's law out to far greater distances than ever before so that you could track the increase in expansion speed against distance far enough away to see if it ever starts to rise more rapidly than merely as speed proportional to distance. Such an upturn would signal acceleration.

It is easy to measure the speed very accurately using the redshift of light but the problem is knowing the distances to the faraway sources of light whose speed you are clocking. If a source appears to be of average brightness, is it because it is intrinsically faint but close by, or is it intrinsically bright but much farther away? Ideally, you want to find a cosmic population of 100-watt light bulbs! You could use your telescope to read the 100-watt label on each one and this would tell you the intrinsic brightness. By comparing that number with the apparent brightness, you could deduce the distance each light bulb was away from us. Alas, there are not populations of labelled light bulbs expanding along with the universe. Instead, one is looking for objects whose intrinsic brightnesses can be determined (like those of the light bulbs) by observing some of their physical properties, like their rate of variation. Such reference objects are referred to as 'standard candles' by astronomers.

These new observations by the two different teams exploited the fact that ground-based telescopes and the Hubble Space Telescope can see exploding stars of a very particular type, called Type 1a Supernovae, out to very great distances. They are good candidates for standard candles because they are believed to result from very particular cosmic events and are some of the brightest objects in the universe. [6]

When stars less massive than 1.4 times the mass of the Sun exhaust their nuclear fuel and implode under their own gravity they contract down to about the size of the Earth, where they can be supported by the counter-pressure created by squeezing the electrons in atoms together.[7] This stable state is called a 'white dwarf' and these stellar corpses are very numerous in the universe. One day, the death throes of our Sun will produce another one.

If a star is heavier, between 1.4 and about three times the mass of the Sun, this electron counter-pressure would be unable to resist the gravitational crushing of the atoms. The electrons would be pushed into the protons in the atomic nuclei and only neutrons would remain. Neutrons also resist being squeezed together and, so long as the mass of the star does not exceed three times that of the Sun, the resisting neutron pressure can stop the gravitational crushing and produce a stable neutron star, just a few kilometres across, with a density 100,000 billion times greater than that of iron. Like white dwarfs, these neutron stars are also very common in the universe, and those that spin rapidly are revealed as pulsars, periodically beaming radiation towards us as

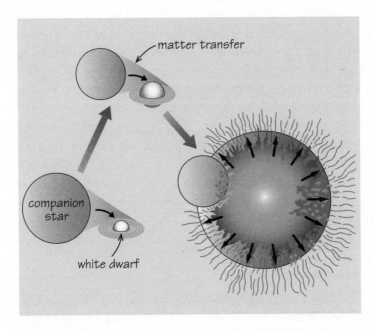

Figure 12.1 A Type Ia supernova occurs when a white dwarf star accretes material from a companion star. Its mass then increases above the Chandrasekhar limit and the white dwarf can no longer support itself against the crush of gravity. The contraction causes a thermonuclear explosion that we witness as a supernova.

they spin like lighthouses. But if the dying star has a mass greater than three times that of the Sun, no known force of Nature can stop it. Eventually, so much mass will fall into so small a region that light will be unable to escape from it. The ongoing contraction will be invisible to the outside universe: a black hole will have formed.

About half of the stars in the universe are in pairs orbiting a common centre of gravity. If one of the stars dies and forms a white dwarf it can keep on accreting matter from the outer regions of its companion star. Eventually, this cannibalism could bring its mass up to just above the 1.4 Sun-mass limit and the electron pressure can no longer support it against the crush of gravity. The white dwarf experiences a dramatic thermonuclear explosion (Figure 12.1). This explosion occurs in every case when the white dwarf's mass just tips over the 1.4 Sun-mass limit and the peak brightness of the explosion will look very similar whenever it happens. That peak brightness is

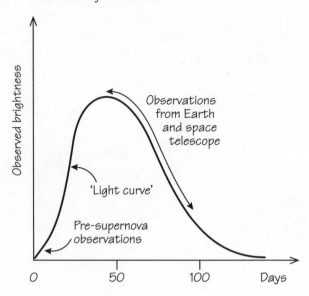

Figure 12.2 A Type Ia supernova light curve. The variation in the observed brightness towards a maximum is followed by a steady return to the light level that existed prior to the explosion. In practice, these variations are tracked in several colours.

more than a billion times that of our Sun – a single star becoming almost as bright as an entire galaxy. After the explosion, the light and its colours fade in a characteristic way over the next few months. This 'light curve' of fading brightness with time is mainly determined by the radioactive decay of the element nickel in the first few days and weeks, and then by the radioactive decay of cobalt. By studying the relation between the peak brightness and the rate of fall-off in the light curve, the two supernova teams compared different super-novae and calculated their relative distances away from us.

The Harvard and Berkeley teams both used this new tool for deter-mining distances and extending our measurements of Hubble's law. They first deployed powerful ground-based telescopes to monitor several hundred patches of the night sky, each containing about a thousand galaxies at the time of the New Moon, when the sky is darkest. Three weeks later they returned and scrutinised the same parts of the sky to see if any of the stars had brightened dramatically into supernovae. Typically, they would find about twenty-five supernovae as they were starting to brighten. Then they followed their light variation using

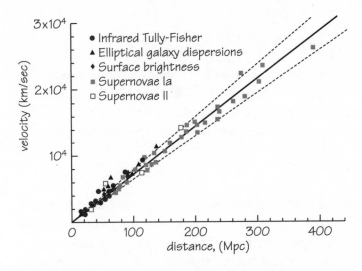

Figure 12.3 Hubble's law of recession velocity versus distance including recent supernovae observations. The graph shows the velocity versus distance observations and the best fit occurs with a slope which gives a Hubble constant of 72 km/sec/Mpc.

ground- and space-based telescopes, watching it grow to its peak and then fall back to pre-explosion levels, monitoring the changes in the light colours as well (Figure 12.2). Remarkably, the light curve shapes they saw were similar to those seen in nearby supernovae of this same type. This gave the observers confidence that the ones they were seeing near the edge of the visible universe were intrinsically the same types of object, and their relative faintness was due solely to their increased distance away from us.

When they put all their data together both groups separately came to the same conclusions. The form of Hubble's law for the expansion speed of the distant supernovae versus their distance away curved upwards (Figure 12.3). The universe was accelerating. This discovery was first announced publicly in January 1998 and has formed a focus of interest for astronomy research ever since, with an ever-growing database of observations, careful study of the two groups' different data sets and analysis techniques, and minute scrutiny of all the assumptions being made about the standard candles and the intervening universe through which the supernova light has travelled before reaching our telescopes.

THE PREPOSTEROUS UNIVERSE

> We have to learn to understand this unattractive universe because
> we have no other choice . . . If I didn't have all of these facts in
> front of me, and you came up with a universe like that, I'd either
> ask what you've been smoking or tell you to stop telling fairy
> tales.
>   John Bahcall

The consequences of these supernova observations were immense.
This was the first direct evidence for the acceleration of the universe.
It confirmed the expectations of the Lambda-CDM cosmological
model quite closely and it proved that the type of anti-gravitating
acceleration that the inflationary-universe theory invoked in the first
moments of the expansion does exist. The simplest possible description of the observed acceleration worked very well. If we just
reinstated Einstein's old cosmological constant as an anti-gravitational
effect and assumed that the expansion was very close to the critical
rate, just as inflation predicted, then we had a universe that was just
like one of Lemaître's that we saw in Figure 3.13, and it matched the
observations very closely.

Although the observations were well explained by the introduction
of a simple cosmological constant term into Einstein's equations,
cosmologists knew that the situation could easily be more complicated. There could be some more exotic anti-gravitating stress in the
universe, like those explored as candidates to drive inflation early in
the universe's history. As a result, the term 'dark energy' was introduced to describe this mysterious energy source. It might always
contribute a constant density to the total density of the universe (like
the cosmological constant) or it might change with time and end up
mimicking the effect of a true cosmological constant in the very
recent past. Remarkably, in order to explain the observations we
require about 72 per cent of the energy in the universe to reside
in this mysterious dark energy and the other 28 per cent to be in
the form of other matter. Of that 28 per cent, just 5 per cent is in the
form of ordinary matter; the rest, 23 per cent, is in the form of some
type of non-atomic cold dark matter whose precise identity is not

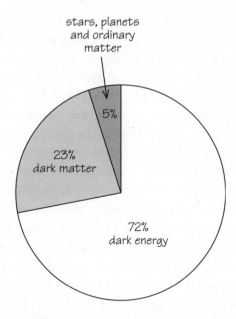

stars, planets
and ordinary
matter

5%

23%
dark matter

72%
dark energy

Figure 12.4 A pie-chart showing the make-up of the universe today. The bulk of the gravitating material is in the form of a 'dark energy' that has been accelerating the universe in the recent past. The remaining material consists of dark matter and luminous matter. It is very likely that the dark matter is predominantly composed of new types of weakly interacting particles, similar to the known neutrinos but much heavier. We hope to confirm their existence in experiments at the Large Hadron Collider in Geneva and then detect them directly with underground detectors when they pass through the Earth.

yet established but is probably a new type of neutrino that could be found soon at the Large Hadron Collider in CERN (Figure 12.4).

Over the past twelve years the case for the acceleration of the universe has grown stronger, with further pieces of indirect evidence supporting it and constraining the details of the dark-energy budget required to drive it. The small fluctuations in the background radiation permeating the universe have been observed in greater detail by a succession of satellite- and ground-based telescopes. The acceleration alters the look-back geometry of the universe (things were smaller at a given time in the past than you thought) and allows us to place limits on how much change can have taken place. This limits the total amount of dark energy and dark matter that the universe can contain. A third new constraint has recently emerged as well. The variations in density that eventually form the clusters and galaxies in the universe should emerge from the hot radiation-dominated era of the universe in the form of huge sound waves. When the temperature falls to about 3000 degrees, electrons are freed from scattering with the radiation and the speed of sound at which the giant acoustic waves travel plummets. The waves retain memory of their size at the moment when that happened. As a result, small ripples were expected to remain in the pattern of clustering shown by the matter in the universe, with a strong additional feature over about 120 megaparsecs imprinted by the residual sound waves. This effect has the rather inelegant name 'baryon acoustic oscillation' because baryons are the collective name for the protons and neutrons that make up the nuclei of ordinary atomic matter (the 5 per cent of the total budget). This gives us new information about the sizes of galaxies at the look-back time when the astronomical survey sees them. If we bring together the observational information from the supernovae, the background radiation and the acoustic oscillations, we get a striking convergence. Fortuitously, the uncertainties and certainties in each of these sets of observations are almost orthogonal, so in combination they narrow the possibilities dramatically. This is evident in Figure 12.5, where the uncertainties (represented by ellipses showing the areas allowed with 68, 95 and 99.7 per cent credibility as they get smaller) are greatly reduced by combining the different pieces of evidence.

The vertical axis shows the fraction of the universe in the form of dark energy. The horizontal axis shows the fraction in all other matter. All the data sets overlap in a narrow region around 0.72 and 0.28 for these two quantities.[8] The overlap also converges around a spatially flat universe with almost zero curvature, which is what we should expect to be if inflation occurred in our past.

This picture is made under the assumption that the dark energy is exactly described by Einstein's cosmological constant. This means it contributes the same density at all times and the vertical axis shows a range of the values that it could have. However, the dark energy might be more exotic – slightly changing its density with time, just like other forms of matter do. This would be the case if the ratio[9] of pressure exerted by the dark energy to its energy density was not equal to -1. When this ratio, w, equals -1 the density is constant and its contribution is exactly the same as the cosmological constant shown in Figure 12.6. If it deviates from -1 there must be a slow change of the dark energy

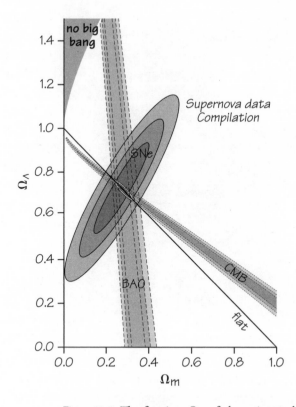

Figure 12.5 The fraction, Ω_Λ, of the universe that can be in dark energy versus the fraction, Ω_m, that can be in other forms of matter, both luminous and dark. The constraints from all observations of supernovae (SNe), the cosmic microwave background radiation temperature fluctuations (CMB), and baryon acoustic oscillation effects in the distribution of matter (BAO) are shown. The three shaded regions show the regions of 68 per cent (darkest shading), 95 per cent (medium-dark shading) and 99.7 per cent (lightest shading) statistical credibility. The overlap region gives the 72 per cent dark energy budget shown in Figure 12.4.

with time. Suppose we allow for this possibility and use all our data once again. Now we can look again where the data converges to agreement inside the 68, 95 and 99.7 per cent credibility contours. The ratio of the pressure to density ratio, *w*, is plotted vertically and the fraction of the universe that is not in the form of dark energy is shown horizontally. If our original assumption about the cosmological constant had been spot on, the data would have converged at the point where *w* equals -1 and matter fraction Ω_m equals 0.28.

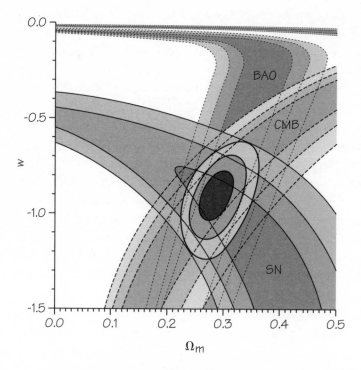

Figure 12.6 The constraints on w, the ratio of the pressure to the density of the dark energy versus the fraction of the universe in matter, ΩM, from observations of supernovae, the microwave background radiation (CMB) and baryon acoustic oscillations (BAO).

At present, the best estimate from all the data is that with 95 per cent confidence the ratio of the pressure to the energy density lies between -1.097 and -0.986. The simplest case of the cosmological constant would require w = -1 and is strongly supported by the observations. As a result the dark energy is very often assumed to be of this form for simplicity.

The situation we find ourselves in today is in many respects very simple but totally mysterious. The visible universe, which may be an infinitesimal part of an infinitely diverse multiverse, is accelerating. It follows the path of the simple model mapped out by Georges Lemaître more than eighty years ago. The geometry of space is very close to being flat and Euclidean and it is seemingly on course to continue expanding for ever. The acceleration is very well described by simply adding a cosmological constant of the sort that Einstein first introduced and then rejected, and which we now interpret as the vacuum energy of the universe, as Lemaître first suggested back in 1934.[10] The resulting best-fit universe has 72 per cent of its energy density in this gravitationally repulsive vacuum form, and the remaining 28 per cent in forms of dark and luminous gravitationally attractive matter. A simple mathematical formula describes how the expansion of the universe develops with time.[11] In Figure 3.13 we saw its characteristic form. Early on, the universe expands as if there is no vacuum energy because it is negligible. As time passes the expansion starts to change gear and shift from deceleration to acceleration when the universe has expanded to about 57 per cent of its present extent. Soon afterwards, the energy density in other forms of matter falls below that of the vacuum energy when the expansion had reached about 73 per cent of its current extent, about 4.5 billion years ago – quite close to the time when the Earth first formed, although no one has suggested that this is anything but a complete coincidence.[12]

One of the most remarkable features of the accelerating universe is its future. In 1986, Frank Tipler and I showed that all information-processing by any type of computer or 'brain' will eventually cease in the future. Star formation or galaxy formation will become extinct. This universe expands too fast for physical processes to keep up. The future universe looks to be a cosmic cemetery of dead stars and isolated elementary particles. And 'isolated' they will certainly be. The acceleration means that there will be a horizon beyond which any observer will be unable to see. It is like being inside a big black hole. Light cannot compete with the acceleration to reach us from very far away. This odd situation was first noticed by Arthur Eddington, back in 1933. In his famous popular book *The Expanding Universe*, he wrote of this phenomenon:

the distance of one galaxy from the next will ultimately become so great, and the mutual recession so rapid, that neither light nor any other causal influence can pass from one to another. All connection between the galaxies will be broken; each will be a self-contained universe uninfluenced by anything outside it. Such disintegration is rather a nightmare to conceive; though it does not threaten any particular disaster to human destiny.[13]

If you are worried about the multiverse introducing unobservable parts of the universe into our picture of the cosmos, then you would also need to worry that even with a 'one-shot' universe there are regions which we cannot see now or even in an infinite future

Yet, despite this good agreement between one of Einstein's simple universes and our observations, there is a deeply disturbing situation here. There are two forms of energy involved in controlling the expansion of the universe over the past 5 billion years. Once the constant vacuum energy takes control, its influence grows stronger and stronger, while that of all forms of matter and radiation diminish rapidly as the universe expands. Tipler and I also showed that this has important consequences.[14] Once the universe starts accelerating, the growth of irregularities caused by the gravitational instability process discovered by Newton, Jeans and Lifshitz switches off. It can't keep pace with the increased rate of expansion. If you haven't formed galaxies by this change-over epoch then you never will. Without galaxies you won't form stable stars and make carbon and the other building blocks of life and observers.

THE PUZZLING UNIVERSE

> . . . the cosmological constant has been like a pair of your grand-
> father's spats – occasionally tried on for costume events – but
> these new results suggest that they are not just coming back into
> fashion, they are now *de rigueur*.
>   Robert Kirshner

The discovery of the existence of a cosmological constant stress in the universe with a value that makes it dominant today was a shock to many astronomers and most particle physicists. If we express the cosmological constant in units of the maximum value it could have then it becomes a pure number, a constant of Nature that characterises how low the vacuum energy of the universe has turned out to be on a scale between zero (the minimum) and one (the maximum).

Until the first direct evidence from supernovae observations appeared, the indirect suggestions that the best fit to all the disparate observations of the background radiation and galaxy clustering needed a cosmological constant didn't convince particle physicists. Their theories led them to expect it existed but the value they expected was 1 on the scale of 0 to 1. Unfortunately, the largest value the observations were allowing before the supernovae data in 1998 was at most about 10^{-120}, about as close to zero as you could imagine. Most particle physicists concluded that it was really zero and there must be some major principle of high-energy physics waiting to be discovered that made it so. When the supernovae observations confirmed that there actually was a cosmological constant in the universe, or something that for all practical purposes is just like it, with this strange tiny value of 10^{-120}, people started scratching their heads. Why this value? What is it that chooses this tiny value which is 10^{120} times smaller than the one that particle physicists believe is the most natural?

The value is doubly, even triply, puzzling because if this tiny value had been about ten times bigger, merely 10^{-119}, no galaxies and stars could have formed. Moreover, if you throw up your arms and say the universe just began with the tiny value of 10^{-120} you don't solve the problem. There is a succession of special epochs during the very early history of the universe when the different forces of Nature peel off and continue with different strengths. Each time this happens, the cosmological constant gets reset to a value that is much larger than the starting value of 10^{-120}. Any physical process that can explain the unusually low value today has to have a long reach forward over cosmic history at different times, to foresee the consequences of different transitions at different epochs and cancel out all the contributions that are coming. We don't know of any physical effects like that.

So far, there is no explanation for this strange value of the cosmological constant that we call the 'dark energy' of the universe. There might be a completely new part of gravitation physics, or its quantum counterpart, that can somehow come up with an explanation. Physicists have started to explore extensions of Einstein's theory of gravity by adding to it new types of geometry so that the curvature of space created by the presence of matter is slightly different and gets *bigger* as the universe gets larger[15]. This type of change was expected when we go back towards the extreme conditions of the initial singularity and the effects of quantum cosmology intervene. But no one really expected to find effects like this changing Einstein's theory at very late times in the universe's history, when the density of matter is very low, gravitational forces are very weak, and there

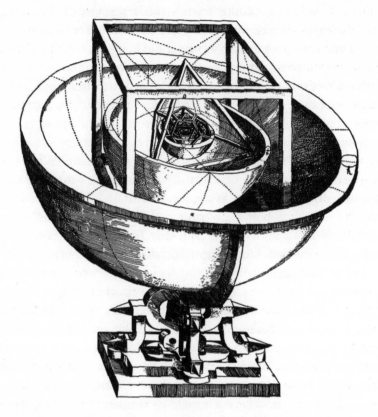

Figure 12.7 Johannes Kepler's deeply Platonic model of the solar system, taken from his first major astronomical work, the *Mysterium Cosmographicum* ('The Cosmographic Mystery') of 1600.

should be no quantum gravitational effects to worry about. Modifications of this sort can certainly be contrived to create late-time acceleration[16] but unfortunately most produce effects elsewhere that we don't see. Those that survive this scrutiny still look just as unnatural as the simple cosmological constant. This is one of the frontiers of modern cosmology.

It is disappointing that, so far, the exploration of string theories and M theory have not helped with this problem at all. Maybe, just maybe, we are looking in the wrong place and there is no explanation for the acceleration of the universe or the cosmological constant of the conventional sort. The cosmological constant may be determined at random by some quantum process in the multiverse landscape. Its value may fall out randomly from universe to universe. We may find ourselves in one of the universes within the multiverse where its value was small enough to permit galaxies and stars to form. While this would be a disappointing outcome for many researchers, we may have to learn to live with it. To those who recoil in horror and think that this would be defeat for the scientific method, I offer a little story, a piece of science faction.

Imagine that in the year 1600 you are seeking to persuade Johannes Kepler that his theory of the solar system need not predict the total number of planets it contains. For Kepler,[17] the number of planets was a manifestation of the deepest mathematical symmetries of Nature. The idea that it would not be determined uniquely and completely by a mathematical model of the solar system would have been anathema, an abnegation of the scientific method.

Kepler had found that each of the five Platonic solids could be uniquely inscribed and circumscribed by spherical balls. If they were nested inside one another like layers on an onion skin the balls would produce six layers, corresponding to the six known planets – Mercury, Venus, Earth, Mars, Jupiter and Saturn. Putting the Platonic solids in the right order – octahedron, icosahedron, dodecahedron, tetrahedron and cube – Kepler found that the nested spheres could be located at intervals corresponding to each planet's (assumed) circular path around the Sun. His drawing is shown in Figure 12.7.

Today, no planetary astronomer in their right mind would try to predict how many planets there should be in the solar system or ascribe any deep significance to their number like Kepler did. That number is just a historical accident, arising from a concatenation of mergers and accidents throughout the solar system's early history that could easily have fallen out differently. It is random outcome. Instead, the planetary astronomer is interested in explaining trends in the composition and sizes of planets with distance from the Sun, and many other detailed aspects of their dynamics.

What if the cosmological constant is a quantity like the number of planets? Just because we think it to be a fundamental property of the universe doesn't mean that it is determined uniquely, completely and inevitably by the laws of Nature. It may be a purely random outcome of those deep laws: an accident, albeit a very important one for us. Nothing more could be said about it. But, just as with the solar system, that isn't the end of the story. There will be other aspects of the visible universe which will be predictable and we must search them out in order to test our theories and explanations. Perhaps the cosmologists of the far future will look back on our concerns about explaining the cosmological constant in the way that we look back on Kepler's preoccupation with the number of planets in the solar system.

But there may be a glimmer of hope. Very recently, Douglas Shaw and I have developed a completely new approach to this problem[18] which adds an additional equation to those first found by Einstein so as to allow the cosmological constant to behave like a changing form of energy if it needs to and to guarantee that a non-quantum picture of the universe is visible at late times. This leads to a nice prediction. If you are an observer in the universe at a time t_u after its expansion begins, you will always observe the effects of a cosmological constant with a value given by $(t_p/t_u)^2$, where $t_p = 10^{-43}$ sec is Planck's fundamental time. Since the present age of the visible universe is $t_u = 4.3 \ 10^{17}$ sec, we expect to see a cosmological constant with a value of about 0.5×10^{-121}, just like we do.[19] Remarkably, we don't require any new forms of dark energy, or quantum modifications of Einstein's equations, or a multiverse of all possible universes in which anthropic selection requires very small values of the cosmological constant. More important still, this new theory makes a definite prediction about the value of the present magnitude of the spatial curvature of the visible universe: this curvature should be positive and the observable value[20] of its effective energy should be equal to -0.0056. Present observations show that this curvature energy lies between -0.0133 and +0.0084 but maps of the microwave background radiation over the sky being made by the European Space Agency's Planck satellite in conjunction with the other observations that contribute to Figure 12.5 will improve the accuracy considerably and we shall soon be able to confirm or refute this very precise prediction.

Our story has come a long way. We started by looking at how hard it is to understand the sky. We saw how difficult it is to make sense of the stars and to decide whether other galaxies exist. Einstein's theory of general relativity in 1915 marked a watershed in our view of our universe and the possibility of others. For the first time we could predict and study whole universes, understand their gross properties and choose between them. We have followed the discovery process that unveiled more and more solutions of Einstein's equations, revealing universes with a host of unexpected properties. We have seen how they were used to make sense of astronomical observations or to couch philosophical views about what the universe should be like.

The current cosmological paradigm is the most long-lived we have seen. It is the inflationary-universe picture, based on a surge of accelerated expansion deep in the universe's past, with its successful prediction of the special pattern of tiny variations we see in the radiation from the early stages of the universe's expansion. The conjunction of super-computer power to simulate the complexities of clustering galaxies with observations of extraordinary sensitivity from new telescopes has mapped out a universe that is understandable but baffling.

We are asked to accept that our universe is a member of a never-ending multiverse of actual universes displaying different properties. Our universe may be special in ways that are essential for our own existence, and that of any other forms of intelligent life. Today, our universe exhibits a second burst of accelerated expansion that began less than 5 billion years ago.

Copernicus taught us that our planet is not at the centre of the universe. Now we may have to accept that even our universe may not be at the centre of the Universe.

Notes

1 Being in the Right Place at the Right Time

1   C. Cotterill, *The Coroner's Lunch*, Quercus, London (2007), p. 123.
2   Presidential Address, Royal Astronomical Society, February 1963, see
     W. H. McCrea, *Quart. J. Roy. Astron. Soc.* 4, 185 (1963).
3   G. Gamow, *My World Line*, Viking, New York (1970), p. 150.
4   The etymology of the word 'universe' can be traced back to the
     use of the Old French *univers*, in the twelfth century, which derives
     from the earlier Latin *universum*. This word is created from *unus*,
     meaning 'one', and *versus*, the past participle of the verb *vertere*,
     meaning 'to turn, rotate, roll or change'. So we have a literal meaning
     of everything 'turned into one' or 'rolled into one'. This usage (with
     the poetic contraction of *universum* to *unvorsum* is found first in
     Lucretius' Latin poem *De rerum natura* ('On the Nature of Things'),
     Bk IV, 262, written in about 50 BC. Another etymology links it
     primarily to the 'everything rotated into one' idea, seeing this as a
     reflection of early Greek cosmologies in which the outer crystalline
     sphere of the heavens rotated and thereby communicated change
     and movement to the planetary spheres inside it, with the Earth
     stationary at their coincident centres.
5   Notably the Irish philosopher Johannes Scotus Eriugena (aka John
     Scotus), AD 815–77. He divided Nature (the most comprehensive
     collection of things) into those which are (have being) and those
     which are not (do not have being). These are then subdivided into
     four classes: (1) those which create and are not created; (2) those
     which are created and create; (3) those which are created and do
     not create; (4) and those which are neither created nor create. Scotus
     placed God in categories (1) and (4), as the beginning and the final
     end of all things. The world of Platonic forms is (2), and (3) is the
     physical world where God acts. In the late thirteenth century the
     medieval church recognised three distinctions in the discussion about

'worlds', or universes. There was the consideration of single worlds
that were sequential in time, worlds that co-existed simultaneously,
and worlds that co-existed in space with empty space in between
them.

6   Genesis 28: 12–13.

7   There is a long tradition of seeking the origin of the world and
ideas about how, or even if, it will end. These began as mythological
adventures that saw the universe as a 'thing' like some of those
things around us on Earth. Like them, it was made out of something
else. Different paradigms emerged, influenced by the way of life of
those who created them. For some, it was a living thing that hatched
out of the activities of the gods. For others, it was caught by a diver
at the bottom of the sea. Others saw it emerge from of a clash of
warring Titans. And some saw it germinating and growing from a
seed, then perishing in its season, before being reborn, cycling periodically between death and rebirth for ever. Some cultures felt no
need for the universe to have begun at all. Its cycle of birth and
death was imagined to be eternal. There was no conception of a
situation in which nothing at all existed: the universe was something
and something could not be nothing. Even the later Christian creation picture of 'creation out of nothing' was no such thing. God
was always there, even when the material universe was not. For
early Greek thinkers like Plato there were always the eternal laws,
or the ideas, behind the appearances. There was no ancient picture
in which the universe came into being without reason or cause. Yet,
this is an unusual idea. We are familiar with all the things and events
around us having causes. My desk has a prior history, a cause that
brought it about from a prior state of less ordered wood. But is the
universe a 'thing' like my desk? Or is it a collection of things like a
society? The distinction might be important because although every
member of a society has a mother – a cause – no society has a
mother. For a wide-ranging look at creation myths in many cultures
see M. Eliade, *The Myth of the Eternal Return*, Pantheon, New York
(1954); M. Leach, *The Beginning: Creation Myths around the World*,
Funk and Wagnalls, New York (1956); C. H. Long, *Alpha: The Myths
of Creation*, George Braziller, New York (1963); E. O. James, *Creation
and Cosmology*, E. J. Brill, Leiden, (1969), and C. Blacker and M.
Loewe (eds), *Ancient Cosmologies*, Allen and Unwin, London (1975).

8   The rotation axis is actually not precisely aligned with the line
through the magnetic North and South Poles.

9   London is at a latitude of 51.5 degrees north and Singapore only
about 1 degree north.

10   The annual path of the Sun traces out a great circle which in ancient times was divided into twelve 'signs' or 'houses' of the zodiac, labelled by the twelve constellations of stars through which the Sun passed on its annual apparent journey around the Earth, which was believed to be at the centre of the celestial motions. Each sign was chosen to cover a region of about 30 degrees in angular extent on the sky (so that the twelve add up to a whole circle of 360 degrees) and about 18 degrees wide (by convention).

11   They are called the northern circumpolar stars.

12   The existence of these unseeable regions, and the fact that they change with latitude and also with time (because the Earth wobbles like a top with a 26,000-year cycle as it rotates on its axis), has led to several attempts to date the creators of the ancient constellations and to ascertain the latitude they observed from. There are gaps in the covering of the sky in the ancient constellation maps which are present because there is a part of the sky whose stars never rise above the observers' horizon. For a recent summary of these analyses, which are fraught with difficulties, see J. D. Barrow, *Cosmic Imagery*, Bodley Head, London (2008), pp. 11–19.

13   For an interesting survey of northern legends associated with a great millstone in the sky, see Hertha von Dechend and Giorgio de Santillana, *Hamlet's Mill*, Gambit, Boston (1969), who argue for a cultural homogeneity in sky myths at northern latitudes created by the image of a millstone in the sky. However, this book has attracted criticism by scholars of archaeoastronomy because of its sweeping generalisations; see the review by C. Payne–Gaposchkin, *J.Hist. Astronomy* 3, 206 (1972).

14   D. G. Saari, *Collisions, Rings, and Other Newtonian N-Body Problems*, American Mathematical Society, Providence, RI (2005).

15   Indeed, for Aristotle the two were logically entwined. In a perfect vacuum there would be no resistance to motion and an infinite speed could result. Both these vetoes, on actual infinities or perfect voids arising anywhere in Nature, were maintained in Western philosophy for more than 1500 years.

16   This was the first time that a topological argument was used in physics. In fact, Aristotle's argument does not uniquely select the sphere as the only shape a rotating Earth could have if it is not to create or enter a void region of space: it could be shaped like any rotational symmetric figure spinning about its centre line.

17   As would any stack of concentric circular discs. But the sphere has this property if you rotate it about any axis.

18   Brian Malow, *Nature*, 11 December 2008.

19   The planets were not attached to the crystal spheres as first proposed, but to circles whose centres were attached to the spheres.

20   Another way in which it can be slightly altered is to move the centre of the orbit of the deferent so that it is slightly displaced from the Earth. This could be done in a different manner for the deferent of each planetary orbit. A further small tuning might be applied by having the orbits occur in slightly different planes.

21   The response, possibly apocryphal, of Alphonso (1221–84) to the mathematical complexities of the Ptolemaic system of planetary motions.

22   In modern studies of celestial dynamics a finite sum of motions like the medieval extension of Ptolemy's epicyclic theory would be called quasi-periodic. If an infinite number of terms are used which converge to a finite sum then the motion is called 'almost periodic' or 'quasi-periodic'.

23   O. Gingerich, *The Book That Nobody Read*, Walker, New York (2004), gives an engaging account of Copernicus's book and its influence. Gingerich has inspected all the known extant copies of Copernicus's book to determine the extent to which they were read, or annotated, and by whom.

24   Such an arrangement had been proposed by Aristarchus of Samos in the third century BC. Writing in *The Sand Reckoner*, Archimedes (287–212 BC) describes this work: 'But Aristarchus of Samos brought out a book consisting of certain hypotheses in which the premises lead to the conclusion that the universe is many times greater than that now so called. His hypotheses are that the stars and the sun remain motionless. That the earth revolves about the sun in the circumference of a circle, the sun lying in the middle of the orbit.'

25   G. B. Riccioli, *Almagestum Novum*, Bologna (1651).

2 The Earnestness of Being Important

1   This general idea is usually known as the Anthropic Principle and was first introduced in an astronomical context by Brandon Carter in a talk to mark the 500th anniversary of the birth of Copernicus in Cracow in 1973. It can be found discussed most extensively in J. D. Barrow and F. J. Tipler, *The Anthropic Cosmological Principle*, Oxford UP, Oxford (1986). More recent applications are discussed in J. D. Barrow, *The Constants of Nature*, Jonathan Cape, London (2003).

2   An observed property of the universe may be an unlikely outcome of a theory in which many possibilities can arise. However, this is

not necessarily a good reason for rejecting the theory. Observers may only be possible in these unlikely universes. Considerations of this sort play an important role in evaluating the predictions of any cosmological theory which possesses an intrinsically random element – so that there are many possible outcomes for different aspects of the universe. In any quantum cosmological theory such a random element is inevitable.

3   The physicist Paul Dirac held this view very strongly.

4   The extra terms in Newton's equations include the so called Coriolis forces, which create the 'Coriolis effect' of the apparent acceleration and deflection of moving objects when observed from a rotating frame, and were first discussed by Gaspard-Gustave Coriolis in 1835. In the same year, Coriolis wrote the definitive study of the game of billiards from the perspective of Newtonian mechanics, *The Mathematical Theory of Spin, Friction and Collision in the Game of Billiards*.

5   The Greek for the Milky Way is *galaxias*, from which we derive the English 'galaxy'.

6   An unpublished manuscript ('Second or Singular Thoughts upon the Theory of the Universe') that was discovered only in the 1960s undid his previous far-sighted conception and replaced it by a universe composed of an infinite number of concentric shells of stars around a divine centre. Various scales of divine punishment were achieved by moving souls between shells that offer greater or lesser degrees of confinement. This model was motivated by alchemical and hermetic concepts of fire and divine action in the creation of the Sun, which are discussed in detail by Simon Schaffer, *J. Hist. Astronomy* 9, 180–200 (1978). The circulation of fire maintained the coherence of the universe and comets were the means by which vital fire is spread around the universe and refuels the Sun.

7   This talk appears in a modern collection of Helmholtz's talks and writings: H. Helmholtz, *Science and Culture: Popular and Philosophical Essays*, ed. D. Cahan, University of Chicago Press, Chicago (1995), p. 18.

8   Up until the middle of the twentieth century, the word 'nebulae' was use to describe fuzzy patches of distant light that astronomers observed. These included both stars and galaxies. Edwin Hubble, for example, referred to galaxies as nebulae. However, today, astronomers reserve the term 'nebulae' to describe regions of dust and gas around stars which are colourfully illuminated by the interactions between their atoms and molecules and radiation streaming outwards from the star. They come in different varieties, for

example planetary nebulae (which have nothing to do with planets!), emission nebulae and reflection nebulae. These spectacular objects are most frequently photographed for astronomical magazines and posters.

9   A little later, the German mathematician Johannes Lambert proposed a similar type of universe that had stars clustered into hierarchies of clusters, and clusters of clusters. Unlike Kant's infinite evolving universe, Lambert's was large but finite and cyclic, with its hierarchy of clustering fanning out like the fractal replications of a repeating pattern.

10 A more technical description is given in Barrow and Tipler, *The Anthropic Cosmological Principle*, p. 620.

11 If v is the rotational speed of a structure moving in a circle at distance r from the centre and the density of matter inside a radius r is $\rho(r)$, then v^2 will be proportional to $\rho(r)\, r^2$.

12 I. Kant, *Universal Natural History and Theory of the Heavens*, transl. W. Hastie, University of Michigan Press, Ann Arbor (1969), p. 149. This volume also contains a translation of the 1751 review of Wright's work which attracted Kant to this subject.

13 Extract quoted from D. Danielson (ed.), *The Book of the Cosmos: A Helix Anthology*, Perseus, New York (2000), p. 271.

14 Kant's view might have been moderated had he known about evolution by natural selection. Our senses are the result of an evolutionary process that selects for correspondence with 'reality' whether we happen to know what it is or not. For instance, our eyes have evolved their structure in response to the real properties of light. The fact that we can understand their structure straightforwardly and accurately in terms of the theory we have about light is telling us that our theory captures much of the essential nature of light that is needed for vision: see J. D. Barrow, *The Artful Universe Expanded*, Oxford University Press, Oxford (2005), pp. 30–33.

15 At that time the retrograde motions of two of the moons of Uranus were not known; they were only discovered by William Herschel in 1798.

16 A. Clerke, *The System of the Stars*, Longmans, Green, London (1890), quoted in Edward Harrison, *Cosmology*, 2nd edn, Cambridge University Press, Cambridge (2000), p. 77.

17 Alun Armstrong (as 'Brian Lane') in *New Tricks*, BBC1, 4 April 2008.

18 A. R. Wallace, *Man's Place in the Universe*, Chapman and Hall, London (1903). Page references are to the 4th edn of 1912.

19 Kelvin was William Thomson until he was ennobled as the 1st Baron Kelvin in 1892. He is buried in Westminster Abbey.

20 He argued that if there were 10 billion stars the velocities would become too large. In gravitating systems containing a total mass M, radius R and average speed of movement v, these three quantities are generally linked by a relation $v^2 \approx 2GM/R$, where G is Newton's constant.

21 Wallace, *Man's Place in the Universe*, p. 248.

22 Ibid., pp. 255 and 261.

23 Ibid., p. 256.

24 Ibid., pp. 256–7.

25 He was particularly struck by the fact that the determination of the speed of light by observations of eclipses of Jupiter's moons matched the value determined terrestrially, concluding that 'These various discoveries give us the certain conviction that the whole material universe is essentially one, both as regards the action of physical and chemical laws, and also in its mechanical relations of form and structure', ibid., p. 154.

26 Ibid., pp. 154–5.

27 W. K. Clifford, *Lectures and Essays*, vol. 1, Macmillan, London (1879), p. 221.

28 S. Brush, *The Kind of Motion We Call Heat*, vols 1 & 2, N. Holland, Amsterdam (1976).

29 The first law of thermodynamics is that energy is conserved.

30 J. Vogt, *Die Kraft*, Haupt & Tischler, Leipzig (1878), and H. Kragh, *Matter and Spirit in the Universe: Scientific and Religious Preludes to Modern Cosmology*, Imperial College Press, London (2004).

31 This motivation can also be found in more recent histories like S. Jaki's *Science and Creation*, Scottish Academic Press, Edinburgh (1974).

32 W. Jevons, *The Principles of Science*, 2nd edn (1877).

33 Note that something which is always increasing (like entropy) did not necessarily ever take a zero value.

34 L. Boltzmann, *Nature* 51, 413 (1895). For a fuller discussion, see Barrow and Tipler, *The Anthropic Cosmological Principle*, pp. 173–8.

35 S. T. Preston, 'On the Possibility of Explaining the Continuance of Life in the Universe Consistent with the Tendency to Temperature-Equilibrium', *Nature* 19, 462 (1879).

36 See, for example the letter to *Nature* from W. Muir, 'Mr Preston on general temperature-equilibrium', *Nature* 20, 6 (1894).

37 The lyric of 'Bye Bye Love' was written by Felice and Boudleaux Bryant and recorded by the Everly Brothers in 1957.

38 Negative curvature surfaces are quite common in living things like vegetable leaves, blossoms and corals because they allow for a larger

surface area to be presented, so the absorption of nutrients is more efficient.

39   Johannes Lambert considered a negatively curved space first in 1786 and Gauss planned surveying experiments to check the geometry of the Earth's surface in 1816. In 1829, Nicolai Lobachevskii wrote a book entitled *The Principles of Geometry*, which showed how new geometries existed in which Euclid's famous 'Fifth Postulate' – that parallel lines never meet – was false but independent of all the other axioms of geometry. The Hungarian mathematician János Bolyai developed similar ideas. Bernhard Riemann subsequently developed the subject into a general description of curved (or 'Riemannian') spaces which he first presented in his doctoral thesis, one of whose examiners was Gauss.

40   Others had also wondered about whether using curved geometries to measure the universe of space might be advantageous. Simon Newcomb had discussed the advantages of a finite spherical space in 'The Philosophy of Hyperspace', *Bull. Amer. Math. Soc.* (2) 4, 187 (1898). For a fuller picture of early interest in non-Euclidean geometry and astronomy, see D. M. Y. Somerville, *Bibliography of Non-Euclidean Geometry*, University of St Andrews and Harrisons & Sons, London (1911).

41   P. H. Harman, *Energy, Force and Matter*, Cambridge University Press, Cambridge (1982).

42   Maxwell and Tait were at school together at Edinburgh Academy and throughout most of their careers vied for honours and prizes in mathematics and physics. Tait is believed to have been the youngest ever senior wrangler (the top student in the Cambridge mathematics course) in 1852, aged twenty. Maxwell came second two years later. Tait is most famous as the mathematician who initiated the theory of knots, while Maxwell is arguably the greatest theoretical physicist after Newton and Einstein.

43   In effect, it would be the three-dimensional surface of an imaginary four-dimensional sphere.

44   In practice, obscuring material might render it unobservable.

45   He showed that the observed Stark effect could be explained by quantum theory and found the most important exact solution of Einstein's general theory of relativity, which describes the gravitational field of a spherical body like the Sun and later turned out to describe the gravity field created by non-rotating black holes. The 'Schwarzschild solution' is a centrepiece of every course in gravitation or astronomy taught anywhere in the world today.

3 Einstein's Universes

1   R. W. Clark, *Einstein: The Life and Times*, World Pub. Co., New York (1971), pp. 385–6.
2   J. Straw, letter to Lord Goldsmith, quoted on BBC website, http://www.bbc.co.uk, on 26 January 2010.
3   J. Eisenstaedt, *The Curious History of Relativity*, Princeton University Press, Princeton (2006), pp. 123–4.
4   We have observed this dragging effect in the vicinity of rotating bodies like the Earth. The effect of masses and motion affects not only the shape of space but also the rate of flow of time. Clocks in strong gravity fields, where space is strongly curved, will be observed to 'tick' more slowly than those far away, where gravity is weak and space almost flat. This effect is also routinely observed in experiments.
5   The most insightful contemporary commentary on this was provided by Erich Kretschmann, who explained how this democratic property that Einstein called 'general covariance' could be arrived at in the formulation of other theories using tensor language: E. Kretschmann, *Annalen der Physik* 53, 575 (1917). For an illuminating discussion of Kretschmann's paper and Einstein's responses to it see R. Rynasiewicz, 'Kretschmann's Analysis of Covariance and Relativity Principles', in H. Goenner, J. Renn, J. Ritter and T. Sauer (eds), *The Expanding Worlds of General Relativity*, Birkhäuser, Boston (1999), pp. 431–62.
6   J. Church, *A Corpse in the Koryo*, Thomas Dunne, New York (2006), p. 266.
7   Quantum mechanics ascribes a wavelike quality to all massive objects. The wavelength is inversely proportional to the mass of the particle. Thus small masses have large wavelength. If that wavelength is bigger than the physical size of the object then it will display overtly quantum properties. But if the mass is very large – like that of you and me – its quantum wavelength will be much smaller than the physical size of the object and it will behave as Newton predicted so long as it moves slowly.
8   N. Coward, *Design for Living*, Act 3, Scene 1 (1933).
9   Tiny changes to Newton's laws or slight asphericity in the shape of the Sun had been suggested.
10 A. Einstein, *Collected Papers of Albert Einstein*, vol. 6, p. 21, original remark made in 1915.
11 Or, equally, one might say that every point is the centre of the surface.

12   If we ask what are the possible force laws for gravity which ensure that the external gravitational effect of a sphere is the same as that of a point of the same mass located at its centre then the $1/r^2$ and r force laws both have this property. The general form of the force law with this property is the one which adds them together in a linear combination $-A/r^2 + Br$, with A and B constants. The total force and induced acceleration are zero when $r^3 = A/B$. Einstein's more elaborate mathematical theory gives the effect when gravity is weak. The constant B is the so-called cosmological constant multiplied by the square of the speed of light.

13   This is $t = 2\pi R/c$ where R is the radius of the space and c is the speed of light.

14   The circumnavigation time in hours is $2/\sqrt{\rho}$ where ρ is the average density in g per cc. The diameter of Einstein's universe is given in terms of its mass, M and the speed of light, c, and Newton's constant of gravitation by $R = 2GM/\pi c^2$. Note that this differs from the formula $R = 2GM/c^2$ for the radius of a Schwarzschild black hole by a factor of π, which arises because of the non-Euclidean spherical geometry of the static universe. This means that its volume is $2\pi R^3$ rather than $4\pi R^3/3$.

15   The Einstein–de Sitter correspondence is detailed and fascinating. A good overview can be found in the article of M. Janssen, https:// netfiles.umn.edu/xythoswfs/webui/_xy-15267477_1-t_ycAqaW0A which provides a commentary on the letters.

16   He called this simply 'solution B', to distinguish it from Einstein's static universe, which he dubbed 'solution A'.

17   A. Eden, *The Search for Christian Doppler*, Springer, New York (1992). This book contains an English translation of Doppler's pioneering paper *On the Coloured Light of the Binary Stars and Some Other Stars of the Heavens*. The dependence of the frequency of sound on the velocity of its source was confirmed experimentally in 1845 by the Dutch physicist Christophorus Buys Ballot, who used a small orchestra of musicians playing a calibrated note on a moving train running on the Utrecht to Amsterdam line.

18   There is much variation in the spelling of his name in the literature: Friedmann, Friedman, Fridman, etc. We will adopt the commonest form, Friedmann, but note that in the published version of his papers in the German journal *Zeitschrift für Physik* his name was printed A. Friedman on the first paper, published in 1922, and A. Friedmann on the second paper, published in 1924. The Russian transliteration of his birth record and the front cover of his book in Russian both have Aleksandr Aleksandrovich Fridman.

19   He was decorated for his work on predicting bomb trajectories, which included flights in aircraft on bombing raids to assess the accuracy of his predictions.

20   There is no evidence to substantiate the story told in George Gamow's book, *My World Line,* that he died as a result of a balloon flight. Gamow had been a student in St Petersburg at the time and was expecting to become a research student of Friedmann's until his death ended that ambition. For a biography of Friedmann, see E. Tropp, V. Frenkel and A. D. Chernin, *Alexander A. Friedmann: The Man Who Made the Universe Expand,* Cambridge University Press, Cambridge (1993).

21   These numbers are very accurate from our modern perspective (the age is 13.7 billion years) and it is not clear why he chose them, or if it was a fortuitous guess. He describes them as figures that 'can only be considered as an illustration of our calculations'.

22   H. Nussbaumer and L. Bieri, *Discovering the Expanding Universe,* Cambridge University Press Cambridge (2009), quoted on p. 90. Friedmann's book is available still in German translation.

23   In his book, Friedmann makes a remarkably insightful point. He points out that although the spherical space of positive curvature creates a universe of finite volume the opposite is not true, the open universe of negative space curvature can have infinite volume but it doesn't have to. That depends on something that the Einstein equations do not control: the topology of space. We will discuss this further in Chapter 11. Friedmann also included Einstein's cosmological constant in his equations.

24   English translations of Friedmann's two papers, 'On the Curvature of Space' and 'On the Possibility of a World with Constant Negative Curvature of Space', together with the two published responses by Einstein, can be found in J. Bernstein and G. Feinberg, *Cosmological Constants,* University of Columbia Press, New York (1989), pp. 49–67, and in the journal *General Relativity and Gravitation* 31, 1991–2000 and 31, 2001–8 (1999).

25   V. A. Fock, *Soviet Physics Uspekhi* 6, 414 (1964), quoted in H. Kragh, *Cosmology and Controversy,* Princeton University Press, Princeton (1996), p. 27.

26   G. Lemaître, *The Primeval Atom: An Essay on Cosmogony,* transl. B. H. and S. A. Korff, Van Nostrand, New York (1950).

27   A Catholic residence for students which later became St Edmund's College, Cambridge.

28   G. Lemaître, 'Un univers homogène de masse constante et de rayon croissant rendant compte de la vitesse radiale des nébuleuses

extragalactiques' ('A Homogeneous Universe of Constant Mass and Growing Radius Accounting for the Radial Velocity of Extragalactic Nebulae'), *Annales de la Société Scientifique de Bruxelles, série A* 47, 49 (1927).

29   My late colleague at the University of Sussex, Sir William McCrea, who knew most of the protagonists from these early years of cosmology quite well, once told me that (Einstein excepted) he always thought of Lemaître as the most powerful scientist of them all, always getting the key results by the simplest methods.

30   This calculation of a Hubble constant equal to 575 km/sec/mpc was not included in the translated version of his paper in 1931.

31   Letter from Hubble to de Sitter 1930, quoted in Nussbaumer and Bieri, *Discovering the Expanding Universe*, pp. 130–31.

32   A. S. Eddington, 'On the Instability of Einstein's Spherical World', *Mon. Not. R. Astron. Soc.* 90, 668 (1930). Einstein visited Eddington in Cambridge around the time that it appeared and so we can assume he heard about it from its author.

33   G. Lemaître, *Mon. Not. R. Astron. Soc.* 91, 483 (1931).

34   Einstein rejected the cosmological constant after he gave up the notion of the static universe, later calling its introduction 'the biggest blunder of my life', yet other cosmologists like Eddington, de Sitter and Lemaître regarded it as an essential ingredient of the theory. Lemaître would go on to show that even if you cast it out of the theory as a new part of the force of gravity, it could always be accommodated as a vacuum energy of the universe. Eddington saw it as a means of connecting the quantum theory and the theories of particles like protons and electrons to the theory of gravity. He claimed that it was the most important ingredient of Einstein's theory. In recent years it has been changed from a theoretical possibility to an observational fact, as we shall see in chapter 11.

35   To a very good approximation you can think of it as being static until some finite time when it begins expanding. The Eddington–Lemaître universe is closed but always expanding from a static state at past infinite time.

36   A. S. Eddington, *The Expanding Universe*, Cambridge University Press, Cambridge, p. 56.

37   For some discussion of this question, see O. Godart and M. Heller, *Pont. Acad. delle Scienze, Commentarii* 3, 1–12 (1929, cited by H. Kragh, *Cosmology and Controversy*), and also see O. Godart and M. Heller, *Cosmology of Lemaître*, Pachart Publ., Tucson (1985). Odon Godart was Lemaître's scientific assistant and Heller, like Lemaître, is both a Catholic priest and a mathematical cosmologist. Ironically, despite

his lack of interest in any form of natural theology or religious apologetic based upon his cosmological ideas, Lemaître found that others used his ideas differently. In 1951, Pope Pius XII made a famous address (not written by Lemaître) about the creation of the universe to the Pontifical Academy, of which Lemaître was the President. He used Lemaître's scientific picture of a universe expanding from a beginning in time as a modern conception of the ancient doctrine of 'creation out of nothing' by a transcendent Deity. Lemaître had tried gently to damp down these confident pontifications during the last few years of his life, but without obvious success.

38 E. R. Harrison, *Mon. Not. R. Astron. Soc.* 137, 69 (1967).

39 These two amusing remarks about their joint paper were made independently to Eddington by Einstein (in discussion) and de Sitter (by letter) in a talk he gave in Cambridge in 1936 and are reported in Eddington's article 'Forty Years of Astronomy' in *Background to Modern Science*, Cambridge University Press, Cambridge (1940).

40 A. Einstein and W. de Sitter, *Proc. Nat. Acad. Sciences* 18, 213 (1932).

41 Distances increase always as the 2/3rd power of the time, $R \propto t^{2/3}$.

42 This critical speed, known as the escape speed of the Earth, is given by $\sqrt{(2GM/R)}$ where M and R are the mass and radius of the Earth respectively.

43 This saying about Newton's laws of gravity first appeared in print in F. A. Pottle, *The Stretchers: The Story of a Hospital Unit on the Western Front*, Yale University Press, New Haven (1929), although it probably originated orally at least fifty years earlier.

44 R. Tolman, *Relativity, Thermodynamics and Cosmology*, Clarendon Press, Oxford (1934), p. 444.

45 J. D. Barrow and M. Dąbrowski, 'Oscillating Universes', *Mon. Not. R. Astr. Soc.*, 275, 850 (1995).

46 The cosmological constant dominates over the attractive part of gravity when the universe gets large enough. The growing entropy makes each cycle get bigger until one gets big enough for the cosmological constant to control the behaviour of the expansion. When that happens, the expansion never reaches a maximum even though the space is closed and it expands for ever, like the Lemaître universes.

47 G. Lemaître, 'The Expanding Universe', *Ann. de la Soc. Scientifique de Bruxelles, série A* 53, 51 (1933). English translation by M. A. H. MacCallum in *Gen. Rel. Gravitation* 29, 641 (1997).

48 R. Tolman, *Proc. Natl. Acad. Sci. USA* 20, 169 (1934).

49 In this extreme situation we say that the initial time slice of the universe across space is 'timelike'. This allows some parts of it to

lie in the causal future of other parts. If this cannot happen then we say that the initial surface is 'spacelike'.

50   E. Hubble, 'Problems of Nebular Research', *Scientific Monthly* 51, 391–408 (November 1940); quotation is from p. 407 and from a public lecture, cited in H. Kragh, *Matter and Spirit in the Universe: Scientific and Religious Preludes to Modern Cosmology*, Imperial College Press, London (2004), p. 153.

51   E. A. Milne, *Z. f. Astrophysik* 6, 29 (1933), reprinted in *Gen. Rel. Gravitation* 32, 1939 (2000); W. H. McCrea and E. A. Milne, *Quart. J. Math. Oxford* 5, 73 (1934), reprinted in *Gen. Rel. Gravitation* 32, 1949 (2000).

52   The scale of the expansion at time t is $R = ct$.

53   The effects of matter and radiation fall off faster than the effect of the negative curvature as time increases and ultimately the universe looks like Milne's special universe. For instance, if the universe had black body radiation added to it then its size would increase like $R^2 = t - kt^2$, where k is the curvature. If k is negative then R is proportional to t when t is large and the expansion looks like Milne's universe. If t is small then R is proportional to the square root of t. Notice also that for the closed universes of positive curvature k is positive (set $k = +1$) and the curve of R (t) is a semi-circle with initial state at $t = 0$, a maximum size at $t = \frac{1}{2}$ and a final state at $t = 1$ time unit.

54   E. A. Milne, *Modern Cosmology and the Christian Idea of God*, Clarendon, Oxford (1952). There is also a section on the subject in his earlier *Relativity, Gravitation and World Structure*, Clarendon, Oxford (1935), p. 138. His views are subtler than might be expected because his cosmological models did not experience a heat death and he maintained that questions about the age, origin, expansion and size of the universe depended upon which system of time-keeping one adopted, so they did not have properties which were independent of the observer being specified. Some observers might, for example, see the constant of gravitation, G, change with time, while others do not. Milne's approach was not closely linked to observations and relied upon principles of uniformity in space and time in order to deduce the model of the universe, rather than build it up combining theory and observations, as we find with Lemaître.

55   Milne died comparatively young, aged fifty-four, of a heart attack in September 1950 while attending a meeting of the Royal Astronomical Society in Dublin.

56   For Milne, the forms of the laws of Nature were necessary truths, like the fact that the angles of a triangle add up to 180 degrees in Euclidean space.

4 Unexpected Universes: the Rococo Period

1   G. K. Chesterton, *Orthodoxy*, London (1908).

2   Take a smooth infinite distribution of matter. Pick any value and direction for the net force of gravity that you want to be acting on you. Now carve out a sphere of matter, with you located on its surface and with the centre in the direction of the force you want acting on you, and an amount of mass inside the sphere to give the magnitude of the force you want acting. Now Newton showed that you only feel the force of gravity from material inside that sphere on whose surface you sit and none from material outside. So you have proved that you feel the force you initially assumed. In this way you can 'prove' that you feel any force you like. This was called the Gravitational Paradox.

3   E. E. Fournier d'Albe, *Two New Worlds: The Infra World. The Supra World*, Longmans, Green & Co., London, (1907). He had written about his idea at a more technical level in his book *The Electron Theory* (1906).

4   E. A. Poe, *Eureka – A Prose Poem*, Putnam, New York (1848). For a discussion of Poe's conception of 'other universes' see E. R. Van Slooten, *Nature* 323, 198 (1986) and A. Cappi, *Quart. J. Roy. Astron. Soc.* 35, 177 (1994).

5   C. Charlier, 'How an Infinite World May Be Built Up', *Arkiv f. Matematik och Fysik* 16, no. 22 (1922).

6   This problem, first noticed by Edmund Halley, was known as Olbers' paradox.

7   For a good review, see J. D. Norton, in *The Expanding Worlds of General Relativity*, Birkhäuser, Boston (1999), pp. 306–8.

8   Selety is a tricky individual to track because he changed his name for some reason, having been Franz Josef Jeiteles up until 1918; see T. Jung, *Acta Historica Astronomiae* 27, 125 (2005). Notice that his new name is just the reverse of his old one and presumably this inversion was to hide his Jewish origins. They are revealed by the –'eles' ending of his original surname. Anti-Semitism was strong in Austria at that time.

9   F. Selety, *Annalen der Physik* 68, 281 (1922).

10 The zero average density arises only in the limiting situation of letting the size become infinite. The density falls off with distance quickly enough for it, when averaged, to be smaller than any number you care to name if you go far enough.

11 The random velocities, v, are related to the density on scale r, $\rho(r)$,

by $v^2 \approx GM/r \propto G\rho(r) \, r^2$ and will not increase with the distance r if the density of matter $\rho(r)$ falls off like $\rho(r) \propto r^{-2}$ or faster. Selety favoured the special situation with $\rho(r) \propto r^{-2}$.

12   A. Einstein, *Annalen der Physik* 69, 436 (1922), with a reply from F. Selety in *Annalen der Physik* 72, 58 (1923).

13   Unfortunately, Einstein seems to have been influenced by an outdated belief that there was no evidence for any external galaxies.

14   The last was in F. Selety, *Annalen der Physik* 73, 291 (1924).

15   G. de Vaucoleurs, *Science* 167, 1203 (2000); J. R. Wertz, *Astrophys. J.* 164, 229 (1971); W. Bonnor, *Mon. Not. R. Astron. Soc.* 159, 261 (1972); C. Dyer, *Mon. Not. R. Astron. Soc.* 189, 189 (1979).

16   For a dissenting opinion, see, for example, F. S. Labini, 'Characterising the Large-Scale Inhomogeneity of the Galaxy Distribution', http://arxiv.org/abs/0910.3833.

17   See his personal account at http://graphics.stanford.edu/~dk/google_name_origin.html.

18   The expansion of distances in three perpendicular x, y and z directions with the time, t, are given by:

$$R_x = t^p, \, R_y = t^q, \, R_z = t^r,$$

where Einstein's equations require that the numbers p, q and r be constrained by the equations

$$p + q + r = 1 \text{ and } p^2 + q^2 + r^2 = 1$$

This means that p, q and r are limited to three non-overlapping ranges of values governed by these inequalities:

$$-1/3 \leq p \leq 0 \leq 1/3 \leq q \leq 2/3 \leq r \leq 1$$

The volume changes as $R_x R_y R_z = t^{p+q+r} = t$ and so increases in time but p is *negative* and so the Kasner universe contracts in the x direction while expanding in the y and z directions.

19   If you try to force Kasner's universe to expand at the same rate in every direction then p, q and r would have to be equal and this is not allowed by the formulae above. In fact, we can make the choice $p = q = 0$ and $r = 1$. This turns out to be the flat space-time of special relativity, or Minkowski spacetime, written in unusual coordinates. This is an allowed solution but is not an expanding universe.

20   This universe was found by Otto Heckmann and Englebert

Schücking in 1959 (*Handbuch der Physik*, vol. 53, Springer, Heidelberg, 1959, pp. 489–519) and has

$$R_x = t^p \,(t + T)^{2/3 \cdot p}, \; R_y = t^q \,(t + T)^{2/3 \cdot q}, \text{ and } R_z = t^r \,(t + T)^{2/3 \cdot r}$$

where T is a constant quantity that can take any value. Roughly speaking, it is the time after which the anisotropic Kasner expansion starts to become increasingly isotropic. We see that when $t \ll T$ the expansion looks like Kasner's universe but when $t \gg T$ it becomes increasingly like the Einstein–de Sitter universe, with $R_x = R_y = R_z = t^{2/3}$. When the cosmological constant, Λ, is added, then $R_x = [\sinh(At)]^p \times ([\sinh(At) + At\cosh(At)]^{2/3 - p}$, where T and $A = (3\Lambda)^{1/2}/2$ are constants. The formulae for R_y and R_z are obtained by replacing p by q and r, respectively. As t approaches zero the Kasner universe is recovered again, but as t grows large we have an approach to a de Sitter universe, with $R_x = R_y = R_z = \exp[t \,(\Lambda/3)^{1/2}]$.

21 Remark of Dirac's to J. Robert Oppenheimer, quoted in G. Farmelo, *The Strangest Man*, Faber, London (2009), p. 121.

22 G. Farmelo, *The Strangest Man*, Faber, London (2009).

23 Quoted in Farmelo, *The Strangest Man*, p. 220.

24 P. A. M. Dirac, *Nature* 139, 323 (1937) and *Proc. Roy. Soc. A* 165, 199 (1938): 'Any two of the very large dimensionless numbers occurring in Nature are connected by a simple mathematical relation, in which the coefficients are of the order of unity.' Eddington's attempts to explain the constants of physics seem to have drawn Dirac's attention to some mysterious numbers that characterise the structure of the universe and the laws of physics.

25 This hypothesis of equality between Large Numbers is not in itself original to Dirac. Eddington and others had written down such relations before, but Eddington had not distinguished between the number of particles in the entire universe – which might be infinite – and the number of particles in the *observable* universe, which is defined to be a sphere about us with radius equal to the speed of light times the present age of the universe.

26 The conclusion $N \propto t^2$ subsequently led Dirac to conclude (P. A. M. Dirac, *Proc. Roy. Soc. A* 333, 403 (1973)), quite wrongly, that this result required the continuous creation of protons. In fact, all it is telling us is that as the universe ages we are able to see more and more protons coming within our visible horizon.

27 Of course, this hypothesis is able to tell us why the different collections of constants N_1, N_2 and $\sqrt{N}$ are of similar magnitude, but not why the magnitude is now close to 10^{40}.

28   The luminosity of the Sun is proportional to G^7 and the radius of
      the Earth's orbit around the Sun is proportional to G^{-1} so the average
      temperature at the Earth's surface is proportional to $G^{9/4} \propto t^{9/4}$.

29   E. Teller, *Phys. Rev.* 73, 801 (1948). Teller, a Hungarian émigré, was
      a high-profile physicist who played an important role in the develop-
      ment of the hydrogen bomb. He and Stan Ulam at Los Alamos
      were the two individuals who came up with the key idea (discovered
      independently by Andrei Sakharov in the Soviet Union and John
      Ward in the United Kingdom) to show how a nuclear bomb could
      be detonated. Later, Teller played a controversial role in the trial of
      Robert Oppenheimer and became an extreme hawk during the Cold
      War period. He was one of the models for the pastiche character
      of Dr Strangelove, so memorably portrayed (much to Teller's annoy-
      ance) by Peter Sellers in Stanley Kubrick's black comedy *Dr
      Strangelove, or, How I Learned to Stop Worrying and Love the Bomb* in
      1964.

30   A change in the value of *e* does not affect the Earth's orbit around
      the Sun, while the luminosity of the Sun is proportional to e^{-6}, so
      the average surface temperature of the Earth is proportional to $t^{3/4}$
      and the era of boiling oceans would be shifted too far into the past
      to be a problem for our biological history.

31   J. Cornwell, *Hitler's Scientists*, Viking, London (2003), pp. 186–90.

32   This was first pointed out by Robert Dicke in 1957 in *Reviews of
      Modern Physics* 29, 363–76, and in *Nature*. It marks a start of so-called
      'anthropic' arguments in cosmology; see J. D. Barrow and F. J. Tipler,
      The Anthropic Cosmological Principle, Oxford University Press, Oxford
      (1986), for a detailed account of these developments.

33   J. D. Barrow, *The Constants of Nature*, Cape, London (2003), p. 111.

34   W. K. Clifford (1876), 'On the Space Theory of Matter,' in *The World
      of Mathematics*, Simon and Schuster, New York (1956), p. 568.

35   Rosen is the R of the famous EPR paradox that was published by
      Einstein, Boris Podolsky and Rosen in 1935 in *Physical Review* 47, 777.

36   D. Kennefick, 'Who's Afraid of the Referee? Einstein and Gravitational
      Waves', http://dafix.uark.edu/~danielk/Physics/Referee.pdf, and
      Physics Today 58 (9), 43-8 (2005).

37   A. Einstein and N. Rosen, *J. Franklin Inst.* 223, 43–54 (1937). Later it
      was appreciated that solutions of Einstein's equations with this form
      had been found much earlier by the mathematician Hans Brinkmann
      in 1925, in H. W. Brinkmann, *Math. Ann.* 18, 119 (1925). Today they
      are called pp waves.

38   R. Feynman, *Surely You're Joking, Mr Feynman!*, Norton, New York
      (1985). G-mu-nu refers to the Einstein tensor or the metric tensor

(one upper case $G_{\mu\nu}$, the other lower case $g_{\mu\nu}$), which appear in Einstein's equations and would be on the lips of anybody talking about general relativity.

5 Something Completely Different

1   R. Goldstein, *Incompleteness: The Proof and Paradox of Kurt Gödel*, W. W. Norton, New York (2005).

2   The sum of the 1st and 100th, 2nd and 99th, etc is always 101. There are fifty of these pairs which sum to 101, so the sum of the first 100 numbers is $50 \times 101 = 5050$. By the same reasoning, the sum of the first N numbers is $N(N+1)/2$. There is a famous story of Karl Friedrich Gauss instantly discovering this when he was nine years old after his schoolteacher thought that he could keep the class occupied for a long time working out this sum the long way.

3   Interestingly, Straus kept his scientific correspondence (thirty-three letters and fifteen manuscripts) with Einstein from this period and his family auctioned the collection at the London Antiquarian Book Fair at Olympia in June 2006 with a reserve price of $1.5 million; see http://www.guardian.co.uk/uk/2006/may/22/science.research.

4   A. Einstein and E. G. Straus, 'The Influence of the Expansion of Space on the Gravitation Fields Surrounding the Individual Stars', *Re. Mod. Physics* 17, 120 (1945), and 18, 148 (1946). The second paper included corrections and additional discussion of the first paper. Straus went on to co-author two further papers with Einstein, in 1946 and 1949, on Einstein's attempts to create a new unified field theory which generalised general relativity still further. After that, all Straus's publications were in number theory.

5   This was the content of Marx's telegram to the Friars Club in Beverly Hills, quoted in *Groucho and Me*, Da Capo Press, New York (1959), p. 321.

6   'The Adventure of the Empty House', in A. Conan Doyle, *The Return of Sherlock Holmes* (1903). This was the first Holmes story set after his supposed demise along with Moriarty at the Reichenbach Falls.

7   For an account of the experiences of Vladimir Fock, one of the leading scientists of the period, who attended lectures by Friedmann as a student and subsequently tried to support research into relativity by steering a difficult course between the ideological obstacles, see G. Gorelik, 'Vladimir Fock: Philosophy of Gravity and Gravity of Philosophy', in *The Attraction of Gravitation*, Birkhäuser, Boston (1993), pp. 308–31.

8   Sadly, he suffered a serious car accident that year which left him unconscious for six weeks, and after which he never regained his mental powers despite living on until 1968. Landau was familiar in detail with the whole of physics and capable of making new discoveries in any part of it.

9   Only forty-three students ever passed this qualifying course to do research with Landau. None completed it more quickly than Lifshitz.

10   He needed to avoid attracting attention in the wrong places and spent 1938–9 in a series of unlikely positions in Moscow and Kharkov, followed by three months with no job at all far away in the Crimea.

11   E. M. Lifshitz, *J. Phys. (USSR)* 10, 116 (1946).

12   The density irregularities behave as if they are closed, denser Friedmann universes embedded in a flat, infinite one. They expand more slowly because they contain denser matter, so the difference in density between the closed universe and the flat background grows.

13   Quoted by Freeman Dyson in J. D. Barrow, P. C. W. Davies and C. L. Harper (eds), *Science and Ultimate Reality*, Cambridge University Press, Cambridge (2004), p. 83.

14   E. Schrödinger, 'The Proper Vibrations of the Expanding Universe', *Physica* 6, 899-912 (1939). Later he extended his study to the Dirac equation for the electron in *Proc. Roy. Irish Acad. A* 46, 25–47 (1940). There is a survey in his two books, *Spacetime Structure*, Cambridge University Press, Cambridge (1950), and *Expanding Universes*, Cambridge University Press, Cambridge (1957).

15   Schrödinger was never fully persuaded of the standard interpretation of the wave function that his equation describes. He stuck to his own view that it represented some type of charge density rather than Born's interpretation of it as a measure of the probability of measuring the occurrence of a possible outcome.

16   S. W. Hawking, 'Black Hole Explosions?', *Nature* 248, 30 (1974).

17   Letter to his mother in October 1961.

18   He was notably paranoid about many things, believing people were trying to poison him and refusing to eat most common foods. When he died in 1978 he weighed 80 pounds and had in effect starved himself to death. His wife Adele acted as a chef, food taster and nursemaid for him and his condition deteriorated significantly after her death in 1970.

19   An earlier solution to Einstein's equations, describing a rotating cylinder of pressure-free material, also contained the possibility of

time travel outside the cylinder because of the severe distortion to space and time caused by the rotation. The solution was found first by the brilliant Hungarian mathematical physicist Cornelius Lanczos (who worked as an assistant to Einstein in 1928–9) in 1924, in *Z. f. Physik* 21, 73, and then rediscovered by the Dutch mathematician Willem van Stockum in 1937, *Proc. Roy. Soc. Edinburgh A* 57, 135, who realised it contained closed time lines. Van Stockum (whose father was the first cousin of Vincent Van Gogh) was a heroic figure. He had been a research student at Edinburgh University and moved to the Princeton Institute of Advanced Study in 1939, hoping to study with Einstein. With the beginning of the war, Van Stockum gave up this ambition and threw himself into the Allied opposition to Hitler, enlisted in the Canadian air force and trained as a bomber pilot, and then joined the Dutch air force in exile in 1944. He became the only Dutch officer to fly missions with RAF Bomber Command and flew many Handley-Page Halifax bombers over Europe. He participated in the D-Day air raids on German gun positions. He died aged just thirty-three, along with his aircrew on 10 June 1944, when his plane was hit by anti-aircraft fire while participating in a 400-plane raid. For more details, see Erwin van Loo's appreciation, 'Willem Jacob Van Stockum: A Scientist in Uniform', June 2004, English translation online at http://www.lorentz.leidenuniv.nl/history/stockum/VliegendeHollander.html.

20   Transcript of radio show interview at http://www.abc.net.au/rn/scienceshow/stories/2006/1807626.htm.

21   K. Gödel, *Reviews of Modern Physics* 21, 447 (1949).

22   M. MacBeath, 'Who was Dr Who's Father?' *Synthese* 51, 397–430 (1982); G. Nerlich, 'Can Time be Finite?' *Pacific Phil. Quart.* 62, 227–39 (1981).

23   Forward-in-time travel, by contrast, is unproblematic and routinely observed in physics experiments. It is what occurs in the so-called 'twin paradox' of relativity theory. One twin goes on a high-speed space trip and returns to find he has aged less than his brother who stayed at home; in effect, the travelling twin has time travelled into the future of the stay-at-home twin.

24   The American philosopher David Malament, in discussing the Grandmother paradoxes, writes of the view that 'time-travel . . . is simply absurd and leads to logical contradictions. You know how the argument goes. If time travel were possible, one could go backward in time and undo the past. One could bring it about that both conditions P and not-P obtain at some point in spacetime. For example, I could go back and kill my earlier infant self, making it

impossible for that earlier self ever to grow up to be me. I simply want to remark that arguments of this type have never seemed convincing to me . . . The problem with these arguments is that they simply do not establish what they are supposed to. To be sure, if I could go back and kill my infant self, some sort of contradiction would arise. But the only conclusion to draw from this is that if I tried to go back and kill my infant self then, for some reason, I would fail. Perhaps I would trip at the last minute. The usual arguments do not establish that time travel is impossible, but only that if it were possible, certain actions could not be performed'; in *Proc. Phil. Science Assoc.* 2, 91 (1984). A distinguished philosopher who has swum against the tide, and argued for the rationality of time travel in the face of the Grandmother paradoxes is the late David Lewis. In 1976, he wrote in his review 'The Paradoxes of Time Travel', *Amer. Phil. Quart.* 13, 15 (1976), that 'Time travel, I maintain, is possible. The paradoxes of time travel are oddities, not impossibilities. They prove only this much, which few would have doubted: that a possible world where time travel took place would be a most strange world, different in fundamental ways from the world we think is ours.'

25   The best limits obtained by Roman Juszkiewicz, David Sonoda and myself in 1985, in the article 'Universal Rotation: How Large Can It Be?', *Mon. Not. Roy. Astr. Soc.* 213, 917 (1985).

6 The Steady Statesmen Come and Go with a Bang

1   F. Hoyle, *The Nature of the Universe*, Blackwell, Oxford (1950), pp. 9–10, based on lectures broadcast on BBC radio in 1949.

2   A. S. Eddington, *The Nature of the Physical World*, Cambridge University Press, Cambridge (1928), p. 85.

3   R. J. Pumphrey and T. Gold, *Nature* 160, 124 (1947); R. J. Pumphrey and T. Gold, *Proc. Roy. Soc. B* 135, 462 (1948); and T. Gold, *Proc. Roy. Soc. B* 135, 492 (1948).

4   This term was introduced by Hong-Yee Chiu in the magazine *Physics Today*, in May 1964, where he wrote that 'So far, the clumsily long name "quasi-stellar radio sources" is used to describe these objects. Because the nature of these objects is entirely unknown, it is hard to prepare a short, appropriate nomenclature for them so that their essential properties are obvious from their name. For convenience, the abbreviated form "quasar" will be used throughout this paper.'

5   H. Bondi and T. Gold, 'The Steady-State Theory of the Homogeneous Expanding Universe', *Mon. Not. Roy. Astron. Soc.* 108, 252 (1948).

6   The expansion scale factor of the de Sitter universe is $a = \exp(H_0 t)$, where H_0 is the constant expansion rate of the universe.

7   F. Hoyle, 'A New Model for the Expanding Universe', *Mon. Not. Roy. Astron. Soc.* 108, 372 (1948).

8   In the 1980s, when the de Sitter universe became the basis for the inflationary-universe theory, this stability property of de Sitter space was rediscovered and recast as the so-called cosmic no-hair theorem.

9   http://www.aip.org/history/cosmology/ideas/ryle-vs-hoyle.htm.

10   For a detailed history of the rivalry between the big-bang and steady-state descriptions of the universe, and the role of astronomical observations in distinguishing between them, see H. Kragh, *Cosmology and Controversy: The Historical Development of Two Theories of the Universe*, Princeton University Press, Princeton, NJ (1996), chapters 4–7.

11   There were of course strong voices in Britain among scientists opposing the fascination with cosmological principles. The most outspoken was Herbert Dingle, one-time President of the Royal Astronomical Society and vociferous opponent of Einstein's special relativity theory, who urged people who had fallen victim to the 'universe mania' to call a 'spade a spade and not a perfect agricultural principle'. For a review of this period of philosophical cosmological argument see 'Cosmology: Methodological Debates in the 1930s and 1940s', *Stanford Encyclopedia of Philosophy*, online at http://www.seop.leeds.ac.uk/entries/cosmology-30s/.

12   Ironically, the conservation of baryon number which preserves the difference between the number of particles and antiparticles carrying baryon charge is no longer believed to be a conserved quantity in Nature. Indeed, it cannot be if the different forces of Nature are truly unified into a 'theory of everything' because that would stop quarks and leptons (like electrons) being able to decay into each other.

13   A. Einstein, *The Meaning of Relativity*, Routledge, London (2003), p. 132.

14   Letter to H. Rood, quoted in H. J. Rood, 'The Remarkable Extragalactic Research of Erik Holmberg', *Publ. Astro. Soc. Pacific* 99, 943 (1987).

15   The connection between this fall-off and the dimension of space was first noticed by the philosopher Immanuel Kant, who pointed out that Newton's inverse-square law of gravity was a consequence of space having three dimensions. If space had n dimensions then these forces would fall off as $1/r^{n-1}$ as the distance, r, increased.

16   Today, we know that encounters and collisions between galaxies

were very common in the past and played an important role in shaping and sizing the wide range of galaxy types that we see in the universe. Holmberg didn't know that but he suspected that close encounters could occur and leave observable effects.

17   Holmberg has simplified things by making the galaxies flat and two-dimensional. He is also assuming that all the interactions take place in this same flat plane (the table-top in his model) as well.

18   E. Holmberg, *Astrophys. J.* 94, 385 (1941). See also Rood, 'The Remarkable Extragalactic Research of Erik Holmberg', p. 921.

19   This development has made the expression and analysis of astronomical data far more visual. The role of pictures and images in astronomy has become extremely important and has also been driven by the evolution of the personal computer. For a discussion of these developments and the whole history of images in science, see J. D. Barrow, *Cosmic Imagery: Key Images in the History of Science*, Bodley Head, London (2008).

20   K. Waterhouse, *The Passing of the Third-Floor Buck*, Michael Joseph, London (1974).

21   Bondi and Lyttleton were unaware of Einstein's suggestion, made in 1924, that this type of charge imbalance might explain the magnetic fields of the Sun and the Earth. Einstein soon gave up the idea after experiments by A. Picard and E. Kessler in 1925 indicated that the charge imbalance was less than $10^{-20}e$, and so ruled out the level of imbalance he required.

22   A. M. Hillas and T. E. Cranshaw, *Nature* 184, 892 (1959).

23   J. G King, *Phys. Rev. Lett.* 5, 562 (1960).

24   Mathematicians might be interested to know that the writer and film producer I. A. L. Diamond (born Domnici in Romania, but his family name was changed in America), who always claimed his initials stood for Interscholastic Algebra League, was an outstanding mathematician at high school and won several gold medals in Mathematics Olympiad competitions in the USA during the period 1936–7. He went on to study at Columbia but gave up his plans for graduate work after his scripts and productions of student revues attracted the attention of Hollywood and they offered him a contract.

25   His autobiography is greatly recommended: G. Gamow, *My World Line: An Informal Autobiography*, Viking, New York (1970), although his autobiographical project was sadly left incomplete because of his death in August 1968.

26   The great Russian cosmologist Yakob Zeldovich (see the memoirs of him by Remo Ruffini at http://arxiv.org/abs/0911.4825, p. 2) said

that Gamow's failure to return to the Soviet Union made him extremely unpopular with all other Soviet physicists because it curtailed any possibility of foreign travel for all of them for a long time afterwards.

27   G. Gamow, *Physical Review* 74, 505–6 (1948).

28   R. A. Alpher and R. Herman, *Nature* 162, 774 (1948).

29   The letter is reproduced in the article by A. A. Penzias, in F. Reines (ed.), *Cosmology, Fusion, and Other Matters*, Colorado Associated University Press, Boulder, pp. 29–47 (1972). It is reproduced here by kind permission of Arno Penzias.

30   R. H. Dicke, *A Scientific Autobiography* (1975), unpublished, held by the National Academy of Sciences.

31   R. A. Alpher and R. Herman, *Physics Today*, 24 August 1988; R. A. Alpher and R. Herman, *Genesis of the Big Bang*, Oxford University Press, Oxford (2001).

32   F. Hoyle and R. J. Tayler, *Nature* 203, 1108 (1964).

7 Universes, Warts and All

1   Remark reported at the British Association for the Advancement of Science meeting in London in 1932, in M. Tabor, *Chaos and Integrability in Nonlinear Dynamics*, Wiley, New York (1989), p. 187. A similar remark is sometimes attributed to Werner Heisenberg on his deathbed in 1976.

2   If universes are allowed to have different properties in different places, then they must be described by partial differential equations rather than ordinary differential equations. That is a huge upgrade in difficulty which presents major problems for computers as well as human calculators.

3   It is for the understanding of solutions of the Navier–Stokes equations. These are the equations which govern fluid flow. They are versions of Newton's famous second law of motion applied to the fluid motion. For a fuller description of the challenge problem, see the Clay Prize web pages at http://www.claymath.org/millennium/.

4   C. F. Von Weizsäcker, *Z. f. Astrophysik* 22, 319 (1943).

5   C. F. Von Weizsäcker, *Naturwissenschaften* 35, 188 (1948).

6   C. F. Von Weizsäcker, *Astrophys. J.* 114, 165 (1951).

7   Heisenberg's visit to Copenhagen in September 1941 to see Niels Bohr has been the subject of much analysis and speculation concerning what they did or did not say to each other about the possibility of building an atomic bomb. This forms the subject

matter of Michael Frayn's drama *Copenhagen* and a historical summary is given by the historian David C. Cassidy, 'A Historical Perspective on Copenhagen', in *Physics Today*, July 2000, pp. 28–32, and in his biography of Heisenberg: *Uncertainty: The Life and Science of Werner Heisenberg*, W. H. Freeman, New York (1992). It seems that during their visit to Copenhagen under the auspices of the German government's Office of Cultural Propaganda Heisenberg and von Weizsäcker gave public talks about the turbulent origins of the solar system.

8   G. Gamow, *Phys. Rev.* 86, 231 (1952).

9   The angular momentum conservation for a spinning eddy of mass M, radius r and rotational speed v requires Mvr to be constant. For ordinary matter M is constant and therefore v is proportional to $1/r$. In the early stages of the universe the dominant energy is in the form of radiation and because of the redshift M is proportional to $1/r$ and so v remains constant. This simple principle also gives the behaviour of spinning perturbations to a uniform expanding universe first found in general relativity by Lifshitz in 1946.

10   This is called the 'inertial range'.

11   The rate of flow of energy is proportional to v^2/t and $v = Lt$ where v is the rotational speed, t is the time and L is the size of the vortex; so eliminating t, we have v^3 proportional to L. This is called the Kolmogorov spectrum.

12   Demonstrating this problem was part of my doctoral thesis at Oxford; see J. D. Barrow, 'The Synthesis of Light Elements in Turbulent Cosmologies', *Mon. Not. Roy. Astron. Soc.* 178, 625 (1977).

13   J. Binney, 'Is the Flattening of Elliptical Galaxies Necessarily Due to Rotation?', *Mon. Not. Roy. Astron. Soc.* 177, 19 (1976).

14   G. Lemaître, *The Primeval Atom*, Van Nostrand, New York (1950).

15   A. Taub, 'Empty Spacetimes Admitting a Three-Parameter Group of Motions', *Annals of Mathematics* 53, 472 (1951).

16   L. Bianchi, *Memorie di Matematica e di Fisica della Societa Italiana delle Scienze, Serie Terza* 11, 267 (1898). For an English translation by Robert Jantzen, see http://www34.homepage.villanova.edu/robert.jantzen/bianchi/#papers.

17   This was the first systematic use of group theory to classify possible universes from their symmetry properties and very different in style (and difficulty) from the mathematical investigations that were being made by other cosmologists at the time.

18   R. B. Partridge and D. T. Wilkinson, *Phys. Rev. Lett.* 18, 557 (1967).

19   'Tous chemins vont à Rome' : Jean de la Fontaine, 'Le Juge arbitre, fable XII, 28, 4' (1693), in Marc Fumaroli (ed.), *La Fontaine: Fables*,

2 vols, Imprimerie Nationale, Paris (1985), or online at http://www.jdlf.com/lesfables/livrexii/lejugearbitrelhospitalieretlesolitaire.

20   C. W. Misner, 'Neutrino Viscosity and the Isotropy of Primordial Blackbody Radiation', *Phys. Rev. Lett.* 19, 533 (1967).

21   Of course, such cosmologists were not as uncritical as Gold makes them sound. They either regarded the question as too difficult to consider at present or suspected there was some other physical principle, as yet undiscovered, that ensured the starting conditions had to be highly symmetrical. Gold was still a strong supporter of the steady-state universe even after the discovery of the microwave background radiation. He knew that in that scenario the present smoothness and isotropy of the universe were the inevitable result of the continuous creation process. Fred Hoyle had successfully proven, together with his research student Jayant Narlikar, that any asymmetries in the steady-state expansion just die quickly away and symmetrical smooth expansion is resumed. The main problem was that this smoothing was so effective that it was hard to understand how there could be any stars and galaxies at all.

22   J. D. Barrow and R. A. Matzner, 'The Homogeneity and Isotropy of the Universe', *Mon. Not. Roy. Astron. Soc.* 181, 719 (1977).

23   http://blog.djmastercourse.com/harmonic-mixing-mixing-in-key/.

24   C. W. Misner, 'Mixmaster Universe', *Phys. Rev. Lett.* 22, 1071 (1969).

25   J. D. Barrow, *Phys. Rev. Lett.* 46, 963 (1981).

26   Manufactured then by Sunbeam Products, which is now a subsidiary of Jarden.

27   Chaos exists because if you compare two Mixmaster universes which are very similar, then they will rapidly become very different after a few of these oscillations. Remarkably, the chaos is described by a deterministic process that corresponds to the expansion of an irrational number into a never-ending continued fraction: V. A. Belinskii, E. M. Lifshitz and I. M. Khalatnikov, *Sov. Phys. Usp.* 13, 745 (1971).

28   The sum of the infinite number of terms, each equal to one-half of its predecessor, in the geometric series $\frac{1}{2} + \frac{1}{4} + \frac{1}{8} + \frac{1}{16} + \ldots$ is 1.

29   C. W. Misner, *Phys. Rev.* 186, 1328 (1969).

30   Quoted by Falconer Madan, in *Oxford outside the Guide-Books*, B. Blackwell, Oxford (1923).

31   J. D. Barrow, P. G. Ferreira and J. Silk, *Phys. Rev. Lett.* 78, 3610 (1997).

32   C. Will, *Was Einstein Right?*, Basic Books, New York (1993).

33   C. Brans and R. H. Dicke, *Physical Review* 124, 925 (1961). A similar type of theory had been developed by Pascual Jordan in 1955 but appeared in book form in German under the title *Schwerkraft und*

Weltall ['The Force of Gravity and the Universe'], Vieweg, Braunschweig (1955), and had not attracted attention, perhaps partly because of the low regard in which Jordan was held after his role in the war. Hoyle was very dismissive of his ideas in his BBC broadcast talks and the book arising from them, *The Nature of the Universe*, Blackwell, Oxford (1950).

34 J. D. Barrow, 'Time-Varying G', *Mon. Not. Roy. Astron. Soc.* 282, 1397 (1996).

35 J. D. Barrow and J. K. Webb, 'Inconstant Constants', *Scientific American*, June 2005, pp. 56–63. For a fuller story see J. D. Barrow, *The Constants of Nature*, Jonathan Cape, London (2002).

36 This is a pure number defined in terms of the charge of an electron, e, the speed of light, c, and Planck's constant, h, by the combination $2\pi e^2/hc$ and is known experimentally to great precision; it is approximately equal to $1/137$. It controls all aspects of atomic and molecular structure.

37 J. D. Bekenstein, *Phys. Rev.* 25, 1527 (1982).

38 H. Sandvik, J. D. Barrow and J. Magueijo, *Phys. Rev. Lett.* 88, 031302 (2002). See also the book by J. Magueijo, *Faster Than Light*, Penguin Books, London (2003).

39 The *Oxford Dictionary of Nursery Rhymes* dates this song to a manuscript compiled between 1770 and 1780.

40 The term 'antimatter' is much older and was coined by the physicist Arthur Schuster in a speculative article, 'Potential Matter: A Holiday Dream', *Nature* 58, 367 (1989), about atoms acting like sources for energy to pour into the universe. It had little to do with the rigorous concept introduced by Dirac in 1928 though.

41 G. Steigman, *Ann. Rev. Astron. Astrophys.* 14, 339 (1983).

42 It was called the conservation of baryon number. Baryon number was a measure of the difference in the number of particles and anti-particles like protons and antiprotons. It was assumed that this difference could not be changed in nature even though the number of particles and the number of anti-particles could change.

43 Y. B. Zeldovich, *Advances Astron. Astrophys.* 3, 241 (1965), and H. Y. Chiu, *Phys. Rev. Lett.* 17, 712 (1966).

8 The Beginning for Beginners

1 A. Einstein, *Sitz. Preuss. Akad. der Wiss. (Berlin)*, pp. 235–7 (1931).
2 H. Robertson, *Science* 76, 221–6 (1932).
3 W. de Sitter, *Mon. Not. R. Astron. Soc.* 93, 628–34 (1933).

4   G. Lemaître, *Publ. Lab d'Astronomie et de Géodésie de l'Université de Louvain* 9, 171–205 (1932).

5   Actually, Kasner's was not an infinite-density event because his universe contained no matter, but the expansion rate and the tidal gravitational forces became infinite. If you introduced matter into Kasner's universe then it did experience infinite density as well, just like in the Heckmann–Schücking universe containing zero pressure matter.

6   The current best estimate for the expansion age of the universe is 13.7 ± 0.1 billion years.

7   E. M. Lifshitz and I. M. Khalatnikov, *Sov. Phys. JETP* 12, 108–13 (1961), 558–63 (1961), and *Advances in Physics* 12, 185–249 (1963); E. Lifshitz, V. V. Sudakov and I. M. Khalatnikov, *Sov. Phys. JETP* 13, 1298–1303 (1961).

8   It was (wrongly) argued that the isotropically expanding universes contain singularities but the isotropically expanding universes are not the most general, hence the most general universes will not contain a singularity.

9   C. W. Misner, *J. Math. Phys.* 4, 924–37 (1963).

10   A. Raychaudhuri, *Physical Review* 98, 1123–26 (1955), *Physical Review* 106, 172–3 (1957). Raychaudhuri's first paper was submitted to the journal in April 1953 but after a long delay was published only in February 1955.

11   A. Komar, *Physical Review* 104, 544 (1956).

12   R. Penrose, *Phys. Rev. Lett.* 14, 57 (1965).

13   This work is described in the book by G. F. R. Ellis and S. W. Hawking, *The Large Scale Structure of Space-Time*, Cambridge University Press, Cambridge (1973).

14   S. W. Hawking and R. Penrose, *Proc. Roy. Soc. London A* 314, 529 (1970).

15   Assumption number 2 about 'no time travel' can be swapped for a different type of assumption if required.

16   In the parlance of mathematics, we say these five assumptions are sufficient (but not necessary) conditions for a singularity to exist in the past.

17   When the gravitational potential reaches values close to the speed of light squared, c^2. Interestingly, it seems that Einstein's theory imposes a limitation on itself: the maximum force that can arise is given by $c^4/4G$, which is approximately 3.25×10^{43} newtons or 10^{39} tonnes.

18   For the history of how this came about, and why, see my book *The Constants of Nature*, Cape, London (2002), chapters 2 and 3.

19   Revelation 3: 16–17.

20   Before the discovery of the microwave background radiation there
      had been some interest in 'cold' universes which were not dominated
      by hot radiation at the beginning. These models were pursued
      briefly in the Soviet Union by Yakob Zeldovich, *Advances in Astronomy
      and Astrophys.* 3, 241 (1965), and in America, at Harvard, by David
      Layzer, *Ann. Rev. Astron. Astrophys.* 2, 241 (1964). Zeldovich quickly
      abandoned the cold model when he understood the new observa-
      tions of Penzias and Wilson in 1965, although Layzer was still arguing
      for a cold early universe, with the background radiation derived
      from thermalised starlight, in 1984; see D. Layzer, Constructing the
      Universe, W. H. Freeman, San Francisco (1984), chapter 8.

21   M. J. Rees, *Phys. Rev. Lett.* 28, 1669 (1972); Y. B. Zeldovich, *Mon. Not.
      Roy. Astron. Soc.* 160, 1P (1972); J. D. Barrow, *Nature* 267, 117 (1977);
      B. J. Carr and M. J. Rees, *Astron. Astrophys.* 61, 705 (1977).

22   Review of the performance of Creston Clarke as King Lear.

23   R. Alpher, J. Folin and R. Herman, *Physical Review* 92, 1347 (1953).

24   Physicists like theories where you can calculate better and better
      estimates which create smaller and smaller changes to the answer
      you last had. These calculations were just the opposite – a sure sign
      that something was badly wrong with the theory being calculated.

25   Traditional cumulative medieval folk song of uncertain origin,
      combining Christian, pagan and astronomical references.

26   H. Georgi and S. Glashow, *Phys. Rev. Lett.* 32, 438 (1974).

9 Brave New Worlds

1.   In the simplest theory of Georgi and Glashow the commonest
      proton decay route was into a positron and a neutral pi meson that
      then decayed very quickly into two high-energy photons. The posi-
      tron would carry 460 MeV of energy and the two photons 240 MeV
      each. The positron energy cascade would occur in the opposite
      direction to that of the two photons that would be separated by
      about 40 degrees.

2   M. R. Krishnaswamy *et al.*, *Phys. Lett. B* 115, 349 (1982).

3   K. Hagiwara *et al.*, *Phys. Rev. D* 66, 010001 (2002).

4   These experiments can be frustrating. In 1996, Super Kamiokande,
      a huge new underground detector containing 45,000 tons of water,
      started operation underground in Japan. It was designed to search
      for proton decay and discover other new neutrino interactions. It
      discovered those neutrino processes very successfully and they led
      to a share in the Nobel Prize for physics. They never found any

evidence for proton decays. Alas, on 12 November 2001 disaster struck the experiment. An accident led to the implosion of the 6600 photomultiplier tubes used to record decay signals. A domino effect was created in which the shock waves from each implosion shattered the neighbouring tubes. Each of them cost about £2000 to replace. The detector was not fully restored to working order until 2006 but is now up and running again.

5   The short list of requirements needed to produce a non-zero matter–antimatter imbalance had been specified in 1966 by the famous Soviet weapons physicist, Nobel Peace Prize laureate and human rights campaigner Andrei Sakharov in a far-sighted paper, *JETP Lett.* 5, 24 (1967).

6   S. W. Weinberg, *Phys. Rev. Lett.* 42, 850 (1979); J. D. Barrow, *Mon. Not. R. Astron. Soc.* 192, 1P (1980); J. Fry, K. Olive and M. S. Turner, *Phys. Rev. D* 22, 2953 (1980).

7   More than nine tenths of the atoms in the universe are hydrogen, whose nucleus contains only a proton. So to very good accuracy, the number of protons is close to the number of atoms in the universe today.

8   R. Browning, 'The Lost Leader', *The Poetical Works of Robert Browning*, ed. G. W. Cooke, Houghton Mifflin, New York (1899), p. 405.

9   Y. B. Zeldovich and Yu. Khlopov, *Phys. Lett. B* 79, 239 (1978); J. Preskill, *Phys. Rev. Lett.* 43, 1365 (1979).

10   Remarkably, in one of his typically beautiful pieces of work, Dirac showed that if just one magnetic monopole existed it would explain why electric charges all come in multiples of a single unit charge: P. Dirac, *Proc. Roy. Soc. A* 133, 60 (1931).

11   One of the oddest episodes in experimental physics was the supposed detection of a magnetic monopole in a detector in the Stanford University Physics Department on 14 February 1982 and thereafter known as the St Valentine's Day monopole. It was never explained and never repeated. Suspicions were high that it was some type of sophisticated practical joke played on the experimenter, Blas Cabrera, because if true it would reflect a general abundance of monopoles at variance with a host of other observations.

12   For a contemporary review, see J. D. Barrow, 'Cosmology and Elementary Particles', *Fundamentals of Cosmic Physics* 8, 83 (1983).

13   *Troilus and Cressida* I, iii, 345.

14   Since they were interested in the very early stages of the universe neither the curvature of space nor the cosmological constant were important and could be assumed to be zero.

15   Alan Guth tells his own story in his book, A. Guth, *The Inflationary Universe*, Addison-Wesley, Reading, Mass (1997).

16   Scalar quantities, like mass or temperature, just have a magnitude that can change with time. Vector quantities, like velocity, have a direction as well as a magnitude.

17   The vacuum state itself looks like Einstein's cosmological constant because this is the only stress that looks the same to all observers, no matter how they are moving. This is necessary for a vacuum state of minimum energy locally because otherwise a lower energy level could be created just by moving relative to it – so it can't be a vacuum.

18   In the cosmological theory without inflation the expansion goes more slowly and it is not possible to 'grow' our whole visible part of the universe from a region that is small enough for light signals to cross it at an early time in the universe's history. Different parts of the region that will expand to become our visible universe therefore end up unco-ordinated, with very different density, temperature and expansion. Specifically, our visible universe is about 10^{27}cm across. Without inflation, if we compress it by a factor of 10^{27} it would be enclosed in a volume 1 cm across. This is the compression corresponding to when the universe was 10^{27} times hotter, and only 10^{-35} seconds old. This seems very small but the distance light has had time to travel in that brief time is only 3×10^{-25} cm, which is vastly smaller than 1 cm. The inflationary acceleration enables our entire visible universe today to have expanded from a region that is smaller than 3×10^{-25} cm in size at this early time.

19   *Macbeth* I, iii, 58.

20   A summary of the workshop was written by J. D. Barrow and M. S. Turner, 'The Inflationary Universe – Birth, Death, and Transfiguration', *Nature* 298, 801 (1982). This long *Nature* article must have set some sort of record as it was submitted in handwritten form and was published five days later. The proceedings of the workshop were published as G. Gibbons, S. W. Hawking and S. T. C. Siklos (eds), *The Very Early Universe*, Cambridge University Press, Cambridge (1983).

21   This was first suggested by Ted Harrison in E. R. Harrison, *Phys. Rev.* (1969), and much used by Ya. Zeldovich, *Mon. Not R. Astron. Soc.* (1972).

22   The latest WMAP technical papers can be found at: http://map.gsfc.nasa.gov/m_mm/pub_papers/threeyear.html.

23   It may be statistically less significant than appears or might be caused by the suppression of very large fluctuations close to the size of the visible universe.

24   It is most probable that our 'bubble' would have inflated by far more than would be necessary to smooth out the region that we can see – if not, there would be a strange and rather anti-Copernican coincidence. This means that it is most likely that our encounter with very different conditions, and even very different physics, in a neighbouring bubble lies in the far, far future after all the stars have died.

25   J. D. Barrow, 'Cosmology: A Matter of All or Nothing', *Astronomy and Geophysics* 43, 4.9–4.15 (2002).

26   Quoted in John Naughton, *The Observer*, Business and Media Section, 18 March 2009.

27   Quoted in the *Sunday Times*, 4 May 2008, p. 15, on his sacking from the Conservative front bench.

28   More specifically, these different possibilities describe the possible 'vacuum states' of the theory. These vacua each possess particular properties which define the physics that will arise if the universe ends up evolving into it, rather than into another one. If the vacuum state had been unique then there would have been only one possible way these fundamental aspects of Nature could have fallen out. In string theory there are extraordinarily many.

29   E. Calabi, *Proceedings of the International Congress of Mathematicians, Amsterdam*, vol. 2, Erven P. Noordhoff/North Holland Publishing, Amsterdam (1954), pp. 206–207; S. T. Yau, *Communications on Pure and Applied Mathematics* 31, 339 (1978). See also B. Greene, *The Fabric of the Universe*, Random House, New York (2004).

30   F. Denef and M. Douglas, *Annals of Physics* 322, 1096–1142 (2007).

31   F. Gmeiner, R. Blumenhagen, G. Honecker, D. Lust and T. Weigand, 'One in a Billion: MSSM-like D-brane Statistics', arXiv:hep-th/0510170.

10 Post-Modern Universes

1   The biggest difficulty is arriving at answers that do not depend on the motion of the observer who deduces them. Einstein's theories do not permit simultaneity to be an absolute, observer-independent concept. Simultaneous for me is not simultaneous for you in the universe. Consequently, trying to evaluate the probabilities of different things arising at the same time in different places suffers from the same ambiguity. Attempts to sidestep this problem by working out all the possible histories that could ensue from one bubble have encountered another problem: that the answers obtained depend upon the vacuum energy of the region assumed

as the starting point for the calculation. A promising new approach has been found which projects the mathematical description of the multiverse on to a higher dimensional 'screen' where the projected sizes of the bubbles that turn into inflated universes can be easily read off. Surprisingly, these two approaches turn out to have strong links, each compensating for the weakness of the other.

2   The general problem of calculating the distribution of probabilities for different outcomes of any process in a universe, or in the whole multiverse, is called the 'Measure Problem'. As a fairly general rule we don't know how to solve this problem for any of the fundamental attributes of the universe.

3   M. Druon, *The Memoirs of Zeus*, Charles Scribner's and Sons, New York (1964).

4   The length of time for which the expansion stayed very close depends on how long the period of inflation lasted.

5   P. A. M. Dirac, 'Reply to R. H. Dicke', *Nature* 192, 441 (1961). In 1980, I received a short handwritten note from Dirac on this subject in which he used exactly the same words that he wrote first in 1938.

6   Quoted in G, Farmelo, *The Strangest Man*, Faber & Faber, London (2009), p.221.

7   J. D. Barrow, 'Life, the Universe, but not quite Everything', *Physics World*, Dec., pp. 31–5 (1999).

8   M. J. Rees, *Comments on Astronomy and Astrophys.* 4, 182 (1972); M. Livio, *Astrophy. J.* 511, 429 (1999).

9   For a detailed survey of the history and content of the idea up until about 1986, see J. D. Barrow and F. J. Tipler, *The Anthropic Cosmological Principle*, Oxford University Press, Oxford (1986), and for recent ideas about the constants of Nature, see J. D. Barrow, *The Constants of Nature*, Jonathan Cape, London (2002).

10   The adjective 'anthropic' was introduced into the subject of cosmology by the British theologian F. R. Tennant in his influential two-volume work *Philosophical Theology*, vol. 2, Cambridge University Press, Cambridge, p. 79, that was published in 1930. Tennant discusses the design of the universe and types of teleology that he imagines could act on a cosmic scale, imagining that the world could be ordered into what he calls 'anthropic categories', which allows it to be selected out of all possibilities and describes the fact that it is consistent with the evolution of intelligent beings. For a fuller discussion of Tennant's work, see J. D. Barrow and F. J. Tipler, *The Anthropic Cosmological Principle*, Oxford University Press, Oxford (1986), section 3.9.

11   B. Carter, 'Large Number Coincidences and the Anthropic Principle in Cosmology', in M.S. Longair (ed.), *Confrontation of Cosmological Theories with Observational Data*, IAU Symposium, Reidel, Dordrecht (1974), p. 132. Carter uses the term 'anthropic principle' here for the first time in modern astronomy; however, it had been used earlier by philosophers.

12   The triangle is the only rigid two-dimensional convex polygon that could be made from struts like this. This is why you find structures like electricity pylons made of arrays of nested triangles and why doubled-barred gates are popular. For more on this see J. D. Barrow, *100 Essential Things You Didn't Know You Didn't Know*, Bodley Head, London (2009), chapter 1.

13   See J. D. Barrow, *The Constants of Nature*, Jonathan Cape, London (2002), chapter 3 for a more detailed account of this correspondence.

14   H. Mankell, *Chronicler of the Winds*, Harvill & Secker, London (2006), p. 25.

15   B. Russell, *Logic and Knowledge*, ed. R. C. Marsh, Allen and Unwin, London (1956).

16   In Voltaire's novel *Candide*, Doctor Pangloss is designed to ridicule this attitude to the world that was enshrined in Leibniz's claim that we lived in the best of all possible worlds. But the obvious criticism that we knew neither what these 'other worlds' were like nor what 'best' might mean was countered by the French mathematician Pierre Maupertuis in 1746. Although Maupertuis was quite critical of the unfounded use of such vague notions by natural theologians seeking to defend the idea that God had made all things for our benefit, he provided a clear-cut meaning to arguments about the best of all worlds. Maupertuis was the first to show that the paths followed by bodies in motion could be determined in two ways. They could be explained by Newton's laws of motion, specifying their starting position and speed and then solving the equations to determine the future path of the body. Alternatively their paths could be found by imagining that they could take all possible paths through space and time between an initial and a final point. The path actually taken is then chosen by requiring that a certain quantity, the 'action', that combines mass, velocity and distance, takes its minimum possible value. This turns out always to give the path followed by Newton's laws. Indeed, the 'least action principle' is a way to derive Newton's laws of motion (and Einstein's as well). Maupertuis argued that the paths of non-minimal action were the 'other' worlds that critics of Leibniz hankered after, and what was meant by 'best' was simply that the action of the path followed in

Nature be the least possible. During the nineteenth century some French commentators even tried to associate new fossil finds with those failed worlds of non-minimal action where life became extinct; see Barrow and Tipler, *The Anthropic Cosmological Principle*, section 3.4, for further details of these developments, which were presented in Maupertuis's book *Essai de cosmologie* in 1750. For the modern philosophical discussion of the concept of possible worlds, see the work of David Lewis on modal realism. Lewis considers all possible worlds to be as real as the actual world because they are the same sorts of thing as actual worlds and cannot be reduced to more basic entities, yet they are causally isolated from each other; see, for example, D. Lewis, *Convention: A Philosophical Study*, Harvard University Press, Cambridge, Mass (1969), and D. Lewis, *Counterfactuals*, Harvard University Press, Cambridge, Mass (1973, rev. edn 1986).

17   A. Guth, in *An Einstein Centenary*, eds S. W. Hawking and W. Israel, Cambridge University Press, Cambridge (1987).

18   E. R. Harrison, 'The Natural Selection of Universes Containing Intelligent Life', *Quart. J. Roy. Astron. Soc. 36*, 193 (1995).

19   Harrison entitled his paper the 'Natural Selection of Universes', although the process he was suggesting is actually akin to the forced breeding of universes.

20   D. Adams, *The Original Hitchhiker Radio Scripts*, ed. G. Perkins, Pan Books, London (1985). This was first broadcast in the episode entitled 'Fit the Seventh' on BBC Radio 4, on 24 December 1979.

21   L. Smolin, *Class. Quantum Gravity 9*, 173 (1992); L. Smolin, *The Life of the Cosmos*, Oxford University Press, Oxford (1996).

22   See, for example, J. A. Wheeler, 'From Relativity to Mutability', in J. Mehra (ed.), *The Physicist's Conception of Nature*, Reidel, Boston (1973), pp. 239 ff., and the last chapter of C. Misner, K. Thorne and J. A. Wheeler, *Gravitation*, W. H. Freeman, San Francisco (1973), p. 1214.

23   Note that the reproduction is very speculative here, unlike in the eternal inflationary universes, where it is the outcome of a definite physical process.

24   More specifically, the gravitational entropy of a black hole, found by Jacob Bekenstein and Stephen Hawking, is proportional to GM^2, so if we ignore shifts in G we get increasing gravitational entropy by increasing M, the mass in black holes, in accord with Smolin's hypothesis. But if G varies too, we can have increasing gravitational entropy with decreasing M if G increases to compensate. Hence it is not natural to suppose that the long-term result of the collapse

and reshuffling of the constants will inevitably lead to a maximum in black hole production.

25   R. Hanson, 'How to Live in a Simulation', *Journal of Evolution and Technology* 7 (2001), http://www.transhumanist.com.

26   J. D. Barrow, *Pi in the Sky: Counting, Thinking and Being*, Oxford University Press, Oxford (1992), chapter 6.

27   N. Bostrom, 'Are You Living in a Computer Simulation?', http://www.simulation-argument.com.

28   However, this is another of those awkward probability measure problems we highlighted earlier in the chapter. Bostrom is implicitly assuming that all the different worlds, fake and real, are of roughly equal probability, or at least not of vastly different probability. This may not be the case.

29   P. C. W. Davies, 'A Brief History of the Multiverse', *The New York Times*, 12 April 2003.

30   L. Susskind, *The Cosmic Landscape: String Theory and the Illusion of Cosmic Design*, Little Brown, New York (2005); A. Vilenkin, *Many Worlds in One: The Search for Other Universes*, Hill and Wang, New York (2006).

31   J. K. Webb, M. Murphy, V. Flambaum, V. Dzuba, J. D. Barrow, C. Churchill, J. Prochaska and A. Wolfe, 'Further Evidence for Cosmological Evolution of the Fine Structure Constant', *Phys. Rev. Lett.* 87, 091301 (2001).

32   S. Wolfram, *A New Kind of Science*, Wolfram Inc., Champaign, Ill. (2002).

33   K. Popper, *Brit. J. Phil. Sci.* 1, 117 and 173 (1950).

34   D. MacKay, *The Clockwork Image*, IVP, London (1974), p.110.

35   J. D. Barrow, *Impossibility*, Oxford University Press, Oxford (1998), chapter 8.

36   Although there is a famous false argument by Herbert Simon claiming the opposite in the much-cited article 'Bandwagon and Underdog Effects in Election Predictions', *Public Opinion Quarterly* 18, Fall, 245 (1954); it is also reprinted in S. Brams, *Paradoxes in Politics*, Free Press, New York (1976), pp. 70–7. The fallacy arose because of the illicit use of the fixed-point theorem of Brouwer in a situation where the variables are discrete rather than continuous; see K. Aubert, 'Spurious Mathematical Modelling', *The Mathematical Intelligencer* 6, 59 (1984), for a detailed explanation.

37   J. D. Barrow, *The Constants of Nature: From Alpha to Omega*, Jonathan Cape, London (2002).

38   In 1965, Gordon Moore, the co-founder of Intel, noticed that every two years the number of transistors that could be miniaturised

and fitted on a square inch of integrated circuit was doubling and the cost halving. This trend has continued to good accuracy and is a reliable predictor of technological progress. Something similar might be true of any civilisation that progresses microtechnologically. The importance of Moore's law for the computer industry was that it enabled software and hardware companies to develop in concert.

39   For a more extended discussion of these arguments and Hume's response, which stimulated Kant's critical work, see Barrow and Tipler, *The Anthropic Cosmological Principle*, chapter 2.

40   D. Hume, *Dialogues Concerning Natural Religion* (1779), in Thomas Hill Green and Thomas Hodge Grose (eds), *David Hume: The Philosophical Works*, vol. 2, London, 1886, pp. 412–16.

41   These questions are closely connected to the issues discussed in Ray Kurzweil's book *The Age of Spiritual Machines*, Viking, New York (1999), concerning the appearance of spiritual and aesthetic qualities in virtual realities and forms of artificial intelligence.

42   Hanson, 'How to Live in a Simulation'.

43   J. L. Borges, *The Library of Babel*, D. Godine, Jaffrey, NH, (2000; first publ. in Spanish, 1941).

44   J. D. Barrow, *The Infinite Book: A Short Guide to the Boundless, Timeless and Endless*, Jonathan Cape, London (2005).

45   This infinite universe must be exhaustively random for the infinite replication 'paradox' to work. It would be no good having a universe that was infinite in volume that contained one atom of matter.

46   F. Nietzsche, *Complete Works*, vol IX, Foulis, Edinburgh (1913), p. 430.

47   The idea of eternal recurrence or return is far older than Augustine. Eudemus of Rhodes, writing in about 350 BC, attributes it to the Pythagoreans: 'If one were to believe the Pythagoreans, with the result that the same individual things will recur, then I shall be talking to you again sitting as you are now, with this pointer in my hand, and everything else will be just as it is now'; G. S. Kirk and J. E. Raven, *The Pre-Socratic Philosophers*, Cambridge University Press, New York (1957), Eudemus Frag. 272.

48   There have been imaginative responses to this dilemma, such as C.S. Lewis's science fiction trilogy *Out of the Silent Planet* (1938), *Perelandra* (1943) and *That Hideous Strength* (1945), which explored the idea that the Earth is a moral pariah in the universe and the other inhabited worlds did not need redemption.

49   G. Ellis and G. B. Brundrit, *Quart. Journal Roy. Astron. Soc.* 20, 37–41 (1979).

50   P. C.W. Davies, *Nature* 273, 336 (1978).

51   Barrow and Tipler, *The Anthropic Cosmological Principle*, chapter 9.5.

52   When Richard Feynman pointed this out to John Wheeler, Wheeler remarked that perhaps this is because there is only one electron.

53   For a recent survey see P. C. W. Davies, *The Eerie Silence: Are We Alone in the Universe*, Allen Lane, London (2010).

54   A Type I civilisation (of which we are an example) can harness the power of a planet, about 10^{17} W of power. A Type II civilisation can harness the power of a star, about 10^{26} W. A Type III civilisation can harness the power of a galaxy, about 10^{37} W. So your civilisation's Type number is roughly given by the formula Type = $[\log_{10}(P) - 6]/10$, where P is the power it can harness in watts. This formula was suggested by Carl Sagan in *Cosmic Connection: An Extraterrestrial Perspective*, ed. J. Agel, Cambridge University Press, Cambridge (1973).

55   J. D. Barrow, *Impossibility*, Oxford University Press, Oxford (1998), pp. 129–31.

56   Denoted as Type I-minus, Type II-minus, etc.: Barrow, *Impossibility*, pp. 133–8.

57   See Barrow and Tipler, *The Anthropic Cosmological Principle*, section 3.8.

58   L. Boltzmann, *Nature* 51, 413 (1895).

59   S. P. Tolver, *Nature* 19, 462 (1879), and *Philosophical Mag.* 10(5), 338 (1880).

60   For another statement of this, see R. Feynman, *The Character of Physical Law*, MIT Press, Cambridge, Mass (1965).

61   R. Penrose, *The Road to Reality*, Jonathan Cape, London (2004).

62   Penrose's argument uses the fact that black holes have a very large quantum entropy associated with their gravitational fields. It is possible to imagine our universe in a state of vastly greater entropy by reorganising the matter into a population of very large black holes. The largest black holes contribute the largest entropy and it is proportional to their surface areas. Their size when the universe has expanded for time t cannot be larger than the speed of light $\times$ t, so their areas and entropies grow as t^2.

63   L. Dyson, M. Kleban and L. Susskind, *J. High Energy Phys.* 0210, 011 (2002); A. Linde, *J. of Cosmology and Astroparticle Phys.* 0701, 022 (2007); D. N. Page, *J. Korean Phys. Soc.* 49, 711 (2006).

11 Fringe Universes

1   O. Stapleton, *Last and First Men*, Penguin Books, Harmondsworth (1972; first publ. 1930), p. 379.

2   *Business Life*, British Airways magazine, October 2007, p. 62.

3   Interestingly, the 'scalar fields' which may be the most dominant forms of energy in the early stages of the universe cannot rotate and their presence would tend to force the universe to be in a state of zero rotation with zero angular momentum.

4   E. Tryon, 'Is the Universe a Vacuum Fluctuation?', *Nature* 396, 246 (1973).

5   See G. Gamow, *My World Line*, Viking, New York (1970), p. 150.

6   The lifetime of the vacuum fluctuation Δt and its energy ΔE will have $\Delta t \times \Delta E \approx h$, where h is Planck's constant of Nature.

7   Some general constraints usually arise but they do not uniquely specify the topology. For example, if the topology is joined up and made finite then all the negatively curved universes of Bianchi and Taub have to expand isotropically: no anisotropic expansion is allowed; see J. D. Barrow and H. Kodama, *Class. Quantum Gravity*, 18, 1753 (2001), and *Int. J. Mod. Phys. D* 10, 785 (2001).

8   G. F. R. Ellis, *Gen. Rel. and Gravitation* 2, 7 (1971).

9   The complete catalogue of possibilities for the universes with negative curvature was an unsolved mathematical problem but the whole list for those with flat geometry was known.

10   D. D. Sokolov and V. F Shvartsman, *Sov. Phys. JETP* 39, 196 (1974).

11   J. R. Gott, *Mon. Not. R. Astron. Soc.* 193, 153 (1980).

12   1 parsec is 3.26 light years, which is about 19 trillion miles or 31 trillion kilometres.

13   Ya. B. Zeldovich and A. A. Starobinsky, *Sov. Astron. Lett.* 10, 135 (1984).

14   In the inflationary scenario it was later found that it is possible to create infinite open universes. In fact, an infinite number of them will appear: A. Linde, *Phys. Rev. D* 58, 083514 (1998), and S. W. Hawking and N. T. Turok, *Phys. Lett. B* 425, 25 (1998).

15   J.-P. Luminet, *The Wraparound Universe*, A. K. Peters, Wellesley, Mass (2008), and J. Levin, *How the Universe Got Its Spots*, Weidenfeld and Nicholson, London (2002).

16   For a lot more detail about this unit of time, called the Planck time after the physicist Max Planck who defined it in 1899, see J. D. Barrow, *The Constants of Nature*, Jonathan Cape, London (2002). The Planck time is defined in terms of the gravitation constant, G, the speed of light in vacuum, c, and Planck quantum constant, h, by the only way of combining these universal constants so as to create a quantity with dimensions of a time: $t_Q = (Gh/c^5)^{1/2}$.

17   B. S. DeWitt, *Phys. Rev.* 160, 1113 (1967); J. A. Wheeler, 'Superspace and the Nature of Quantum Geometrodynamics', in C. D. DeWitt

and J. W. Wheeler (eds), *Battelle Rencontres: 1967 Lectures in Mathematics and Physics*, Benjamin, New York (1968), p. 242.

18   J. Hartle and S. W. Hawking, *Phys. Rev. D* 28, 2960 (1983).

19   S. W. Hawking, in H. A. Bruck, G. V. Coyne and M. S. Longair (eds), *Astrophysical Cosmology*, Pontifical Academy, Vatican (1982).

20   At first, it was thought that the prescription for this type of quantum universe required space to be finite in the universe. Soon, however, Hawking and Neil Turok at Cambridge realised that it was quite possible to accommodate infinitely large spaces as well.

21   Sometimes this is described as quantum 'tunnelling' from nothing. It refers to the property of quantum mechanics which permits transitions to occur which are forbidden in Newtonian mechanics. This is called quantum tunnelling. The allusion is to the idea that if you don't have enough energy to climb up and over a hill to get to the other side you might be able to tunnel through the middle.

22   The situation was for some time not so clear-cut, as Vilenkin initially made an error in his presentation which made it appear that his boundary condition gave rise to the same expectations as Hartle and Hawking's. Soon several people noticed the mistake and it was corrected by Vilenkin in 1984 in *Nucl. Phys. B* 252, 141 (1985). For Vilenkin's commentary on all these developments, see the account in A. Vilenkin, *Many Worlds in One*, Hill and Wang, New York (2006). Then the real difference between the two proposals became clear.

23   A. Vilenkin, *Phys. Lett. B* 117, 25 (1982).

24   J. D. Barrow, in B. Carter and J. Hartle (eds), *Gravitation in Astrophysics*, Plenum, NATO Physics series B, vol. 156, p. 240.

25   J. R. Gott and L.-X. Li, 'Can the Universe Create itself?', *Phys. Rev. D* 58, 023501 (1998).

26   Letter to the Editor, *The Independent* newspaper, 8 July 2004, p. 28.

27   J. D. Barrow and M. Dąbrowski, *Mon. Not. Roy. Astron. Soc.* 275, 850 (1995).

28   J. Khoury, B. Ovrut, P. Steinhardt and N. Turok, *Phys. Rev. D* 64, 123522 (2001). For a popular account, see P. J. Steinhardt and N. Turok, *Endless Universe: Beyond the Big Bang*, Doubleday, New York (2007).

29   The terminology 'brane' and 'braneworlds' comes from generalising a membrane to more dimensions than two. The terminology is a source of endless amusement – universes on the brane, p-dimensional braneworlds are known as p-branes (we are talking about the $p = 3$ case here), removing sections from the spaces would be brane surgery, I presume, etc.

30   The model avoids a continual build-up of radiation entropy from cycle to cycle because it undergoes accelerated expansion in each cycle and this dilutes the entropy far more in the expansion phase than in the ensuing contraction.

31   Initially it was thought that these all die away very fast, but when the role of collisionless particles was included this conclusion didn't remain: J. D. Barrow and K. Yamamoto, *Phys. Rev. D* 82, 063516 (2010).

32   M. P. Salem, 'Bands in the Sky from Anisotropic Bubble Collisions', http://arxiv.org/PS_cache/arxiv/pdf/1005/1005.5311v1.pdf.

33   D. Thomas, *Selected Poems*, ed. W. Davies, J. M. Dent & Sons, London (1974), p. 131. This poem was for the poet's father as he approached blindness and death.

34   A. Albrecht and J. Magueijo, *Phys. Rev. D* 59, 043516 (1999); J. D. Barrow, *Phys. Rev. D* 59, 043515 (1998). A broader story about this work is told in the book by João Magueijo, *Faster Than Light: The Story of a Scientific Speculation*, Penguin Books, London (2003).

35   Inflation solves the flatness problem by making the density terms fall off more slowly than those controlling the curvature so the curvature becomes negligible after much expansion has occurred. Varying speed of light does the opposite: it makes the curvature and cosmological constant terms fall off faster than the density terms so again the universe becomes flatter after much expansion has occurred.

36   J. D. Barrow and J. Magueijo, *Phys. Lett. B* 447, 246 (1999).

37   J. Magueijo and J. Noller, *Phys. Rev. D* 81, 043509 (2010).

38   See my earlier book, Barrow, *The Constants of Nature*, for a fuller account of this search.

39   J. D. Barrow and J. K. Webb, 'Inconstant Constants', *Scientific American*, 55–63, June 2005.

12 The Runaway Universe

1   Quoted in C. Brownlee, 'Hubble's Guide to the Expanding Universe', National Academy of Sciences Classics, online at http://www.pnas.org/misc/classics2.shtml.

2   C. Green, *The Human Evasion*, Hamish Hamilton, London (1969), p. 12.

3   The conference proceedings can be read in the volume *Critical Dialogues in Cosmology*, ed. N. Turok, World Scientific, Singapore (1997).

4 M. Davis, G. Efstathiou, C. Frenk and S. D. M. White, *Astrophys. J.* 292, 371 (1985).

5 A. Riess *et al.*, *Astron. J.* 116, 1009 (1998), and S. Perlmutter *et al.*, *Astrophys. J.*, 517, 565 (1999).

6 Using these supernovae as standard candles was suggested by David Arnett in 1979 although for a long time people thought it would be too difficult to do so in practice.

7 The pressure is generated quantum-mechanically by the resistance that electrons have to being put into the same state and is called electron degeneracy pressure. The resulting white dwarf states have atoms and their electrons packed as closely as is allowed.

8 The most precise estimate from this data is currently 0.721 ± 0.015 and 0.279 ± 0.015 with 68 per cent confidence if the geometry of the universe is flat.

9 This is just the simplest possible relationship between the pressure and the energy density. There are many more complicated possibilities which have been explored in detail.

10 G. Lemaître, *Proc. Nat. Acad. Sci.* 20, 12 (1934).

11 The equation $a(t) = a_0 \sinh^{2/3}(t\sqrt{3\Lambda}/2)$ describes the curve in Figure 3.13 where a_0 is the present value of the expansion radius, $a(t)$, and t is the time since the expansion began.

12 This coincidence was first noticed by Eddington in his paper on the instability of Einstein's static universe, *Mon. Not. Roy. Astron. Soc.* 90, 677 (1930). He remarks that we should count ourselves 'extraordinarily fortunate' to have arrived on the cosmic scene 'just in time to observe this interesting but evanescent feature of the sky'.

13 A. S. Eddington, *The Expanding Universe*, Cambridge University Press, Cambridge (1933), pp. 85–6. Eddington called the separated parts of the accelerating universe 'bubbles' and defined them as 'regions between which no causal influence can ever pass'. In the epigraph heading this chapter of that book, Eddington quotes Francis Bacon's famous poem 'The Life of Man': 'The world's a bubble, and the life of man / Less than a span'. This seems to be the source of his 'bubble' terminology.

14 J. D. Barrow and F. J. Tipler, *The Anthropic Cosmological Principle*, Oxford University Press, Oxford (1986), section 6.9.

15 S. M. Carroll, V. Duvvuri, M. Trodden and M. S. Turner, *Phys. Rev. D* 70, 043528 (2004).

16 There is a simple condition that shows when a generalisation of Einstein's theory will have de Sitter's accelerating universe as a solution. It was found by J. D. Barrow and A. Ottewill, *J. Phys. A* 16, 2757 (1983). In fact, most of the generalisations of Einstein's theory

considered to generate the acceleration can be shown (see J. D. Barrow and S. Cotsakis, *Phys. Lett. B* 214, 515 (1988), where this is proved) to be disguised versions of Einstein's theory to which other forms of matter have been added. Interestingly, they are the same types of matter that had been appealed to in order to create the phenomenon of inflation early in the universe's history.

17   M. Caspar, Kepler, transl. C. D. Hellman, Dover, New York (1993); O. Gingerich, *The Eye of Heaven: Ptolemy, Copernicus, Kepler*, Springer, New York (1997).

18   J. D. Barrow and D. J. Shaw, A New Solution of the Cosmological Constant Problems (2010), arXiv gr-qc/0073086; D. J. Shaw and J. D. Barrow, A Testable Solution of the Cosmological Constant and Coincidence Problems (2010), arXiv gr-qc/10104262.

19   The appearance of the present age of the universe in this prediction arises because the only contributions to the value of the cosmological constant's value are made by sources which are in causal contact with us, that is roughly closer than a distance ct_u.

20   This quantity is the fraction of the energy density of the universe residing in the gravitational effects of curvature today and is defined to be the value of $\Omega_k = -k/a^2H^2$ today. It is analogous to the quantities plotted in Figure 12.5 for the fractional contributions to the energy density of the universe arising from the matter and the cosmological constant. Note that our prediction that $\Omega_k = -0.0056$ is *negative* requires $k > 0$.

Picture Credits

Fig. 1.1 Reproduced from J. D. Barrow, *The Artful Universe*, Clarendon Press, Oxford (1995). ● Fig. 1.2 Reproduced from Archie E. Roy, 'The Origin of the Constellations', *Vistas in Astronomy*, 27, 171–97. Copyright © 1984, reprinted by permission of Elsevier. ● Fig. 1.4 Herman Eisenbeiss/Science Photo Library. ● Fig. 1.6a Henry E. Huntington Library and Art Gallery. ● Fig. 1.9 Royal Astronomical Society/Science Photo Library; from Nicolaus Copernicus, *De revolutionibus orbium coelestium* (1543). ● Fig. 1.10 Reproduced from E. R. Harrison, *Cosmology: The Science of the Universe*, Cambridge University Press, Cambridge (1981). ● Fig. 2.2 Reproduced from E. R. Harrison, *Cosmology: The Science of the Universe*, Cambridge University Press, Cambridge (1981). ● Fig. 2.3 akg-images. ● Fig. 2.4 Reproduced from E. R. Harrison, *Cosmology: The Science of the Universe*, Cambridge University Press, Cambridge (1981). ● Fig. 2.5 Reproduced from A. R. Wallace, *Man's Place in the Universe*, Chapman and Hall, London (1912). ● Fig. 2.7b Lynne McDonagh. ● Fig. 2.9 Archive, Astrophysikalisches Institut Potsdam (AIP). ● Fig 3.2 Courtesy of the Albert Einstein Archives, The Hebrew University of Jerusalem, Israel. ● Fig. 3.9 akg-images. ● Fig. 3.12 Reproduced from A. V. Douglas, *The Life of Arthur Stanley Eddington*, Nelson, London (1956). ● Fig. 3.14 Science Photo Library. ● Fig. 3.15 Leiden Observatory Archives; New York Times Photo Archives/REDUX/Eyevine. ● Fig. 4.1 Mary Lea Shane Archives/Lick Observatory. ● Reproduced from E. R. Harrison, *Cosmology: The Science of the Universe*, Cambridge University Press, Cambridge (1981). ● Fig. 4.3 Pasieka/Science Photo Library. ● Fig. 4.4. Courtesy of University Archives, Columbia University in the City of New York. ● Fig. 4.6 Roger Viollet/Getty Images. ● Fig. 4.7 Markus Pössel (MPIA and AEI)/http://www.einstein-online.info. ● Fig. 5.1 Courtesy of Professor Daniel Straus. ● Fig. 5.2b Charles Dyer and Allen Attard. ● Fig. 5.4 John D. Barrow. ● Fig. 5.6 Time & Life Pictures/Getty Images. ● Fig. 6.1 Reproduced by permission of the Master and Fellows of St John's College, Cambridge. ● Fig. 6.2 American Astronomical Society/NASA Astrophysics Data System. Reproduced from *Astrophysical Journal*, vol. 94, no. 3, 1941. ● Fig. 6.3 American Astronomical Society/NASA Astrophysics

Data System. Reproduced from *Astrophysical Journal*, vol. 94, no. 3, 1941. ● **Fig. 6.5** Reproduced courtesy of Dr Arno Penzias. ● **Fig. 6.6** Cover of R. A. Alpher's *Genesis of the Big Bang*, Oxford University Press, Oxford (2001); cover design by Emily Kulp. ● **Fig. 6.7** Science Photo Library. Reproduced from R. V. Wagoner, W. A. Fowler and F. Hoyle, 'On the Synthesis of Elements at Very High Temperatures', *Astrophysical Journal*, vol. 148, no. 3, 1967. ● **Fig. 7.1** AirTeamImages, copyright © Steve Morris 2010. ● **Fig. 9.9** Reproduced from M. Lachièze-Rey, J.-P. Luminet and J. Laredo, *Celestial Territory: From the Music of the Spheres to the Conquest of Space*, Cambridge University Press, Cambridge (2001) ● **Fig. 9.10** Copyright © ARS, NY and DACS, London 2010; image copyright © The Metropolitan Museum of Art. ● **Fig. 11.1** Mehau Kulyk/Science Photo Library. ● **Fig. 11.3** Emilio Segrè Visual Archives/ American Institute of Physics/Science Photo Library. ● **Fig. 11.4** Reproduced courtesy of Kip S. Thorne.

Index

Note: Figures are indicated by *italic page numbers*, end notes by suffix 'n' (e.g. 298n[24] = note (24) on page 298)